WHERE TO WATCH BIRDS IN PORTUGAL, THE AZORES & MADEIRA ARCHIPELAGOS

Published by Pelagic Publishing
www.pelagicpublishing.com
PO Box 874, Exeter, EX3 9BR, UK

Where to Watch Birds in Portugal, the Azores & Madeira Archipelagos
ISBN 978-1-78427-223-4

Reprinted 2020

This book is a reprint (without changes) of *The Prion Birdwatchers' Guide to Portugal, the Azores & Madeira Archipelagos* (2nd Edition, 2014, ISBN: 978-1-87110-413-4) published by Prion Ltd, Cley, UK.

British Library Cataloguing in Publication Data
A catalogue record for this book is available from the British Library.

Cover photograph: Little Bustard *Tetrax tetrax* adult male in courtship display. Sylvain Cordier/Biosphoto/FLPA

WHERE TO WATCH BIRDS IN PORTUGAL, THE AZORES & MADEIRA ARCHIPELAGOS

COLM MOORE, GONÇALO ELIAS
AND HELDER COSTA

Illustrations by Tony Disley & Martin Woodcock

Pelagic Publishing

ABOUT THE AUTHORS

Colm Moore, formerly head of science at one of Lisbon's international schools, lived more than a quarter of his life in Portugal before moving to Beijing, China where he continues to pursue his career. He is a founder member of the Portuguese Rarities Committee and has published more than forty papers and three books on a variety of subjects from lichen ecology to bird topography.

Gonçalo Elias, has devoted almost thirty years to birding activities, and has been actively ringing birds since 1992. He is author and coauthor of several books and numerous papers on the birds of Portugal. He is the founder and manager of the avesdeportugal.info portal, the most comprehensive website about the birds of Portugal, online since January 2008. He has a degree in telecommunications engineering and an MBA, and he is a professional trainer accredited by the Portuguese Institute for Employment and Vocational Training.

Helder Costa, when not engrossed in the banking business, is one of Portugal's leading ornithologists. Over the last 30 years he has travelled all over the country collecting an enormous amount of information about birds and places to watch them. He was a founder member of the Portuguese Rarities Committee and a past President of Sociedade Portuguesa para o Estudo das Aves (SPEA – the Portuguese Society for the Study of Birds). He is author or co-author of 13 books and more than 60 papers.

CONTENTS

ACKNOWLEDGEMENTS

While preparing the second edition of this book we were lucky enough to have contributions from some experienced birdwatchers who either commented upon early drafts of the texts or provided very useful information about certain regions and sites.

Filipe Bally provided invaluable advice on the text about Ria de Alvor. Thijs Valkenburg made useful comments about the Algarve section and the Full Species List. João Nunes provided insightful material about Madeira as did Carlos Pereira for the Azores. Last but not least, Romão Machado gave us excellent support, heading an 'expedition' to Ria de Aveiro and the surrounding areas.

We would like to express our gratitude to them all while emphasising that any errors or omissions which may be found in this book are the responsibility of the three authors alone.

INTRODUCTION

Portugal is a beguiling challenge for birdwatchers. It has the temperate forests of western Europe, and the continental plains of the east; it is washed by a cold enriching current yet is climatically dominated by a subtropical anticyclone.

The country lies at the south-western corner of Europe and despite facing the Atlantic Ocean it is subject to a strong Mediterranean influence, both climatically and economically. It borders only one other country (Spain) and Portugal's history is strongly connected to that of its neighbour.

Apart from the mainland rectangle which forms most of the western part of Iberia, Portuguese territory includes three archipelagos in the Atlantic which are a large part of the Macaronesian region. These archipelagos include the Azores, which lie halfway between Europe and North America; Madeira, lying west of the Moroccan coast; and the remote uninhabited Selvagens Islands, which lie not far from the Canary Islands but politically belong to Portugal. All are of volcanic origin.

For anyone of an enquiring mind, that natural caprice called Portugal is fascinating. This small sliver of land in the corner of Europe has certainly had its moments. Home to a highly original culture in the fourth and third millennia BC, wintering quarters to Carthaginian armies and a sphere of Roman influence by 60 BC, Portugal embraced Judeo-Christianity and Islamic fervour and gained immeasurably by the cultural aftershock. Though the Portuguese Inquisition was terminated only in 1820, the two great dynasties that ruled Portugal from early in the 12th century to 1580 recognised Moorish and Jewish genius and thereby fulfilled Portugal's dream by discovering the world, often using regal prerogative to side-step church intolerance.

Colonial Portugal was immense yet it might have gone the way of peaceful transition after World War II. However, Oliveira Salazar, leading a country 'stopped in time and history', propelled Portugal into a colonial war. Yet that subtle mix of race, the Portuguese, a people slow to burn, pragmatic and contradictory, brought about a bloodless revolution in 1974 at home and disengagement abroad.

In 1986 Portugal became an active member of Europe's latest alliance, the European Union, and has used its colonial experience to forge new links with lusophone Africa and South America. Shortly after the new millennium began, the Euro (€) emerged as the main currency in continental Europe and Portugal joined the club, yet this option has led to economic stagnation and ultimately to a sovereign debt crisis that has pushed the country into recession.

It is widely believed that tourism, including bird-related tourism, may have an important role to play in restoring the country's economic situation to something approaching normality. Birdwatching as a tourist attraction in Portugal has undergone a marked development in recent years and conditions have improved for the visiting birdwatcher, both access-wise and information-wise making it much more birdwatcher-friendly than it was two decades ago.

BRIEF NOTES ON PORTUGUESE ORNITHOLOGY

The evolution of ornithology and birdwatching in Portugal is a story of great beginnings and periods of inactivity. From a golden period in the late 19th and early 20th centuries, when the first detailed works about Portuguese birds were produced, there followed a 'dark age' from the mid-1930s until the 1960s. However, by the late '60s, Sociedade Portuguesa de Ornitologia (SPO) was up and running in Porto, the first step towards a modern approach to birding. SPO lasted until the 1980s and by then *Centro de Estudos de Migrações e Protecção de Aves* (CEMPA), a state body, had emerged and started to develop regular waterbird censuses and ringing activities. In 1989, CEMPA published the first breeding bird atlas covering the mainland territory of Portugal.

In 1993, Sociedade Portuguesa para o Estudo das Aves (SPEA), a non-governmental organization primarily concerned with birds and conservation, was founded and was a keystone endeavour in developing ornithology in the country. One of the first big achievements of the new society was the creation of the Portuguese Rarities Committee (PRC) which appeared in 1995. Since then there has been a huge increase of interest and public awareness about birds and their habitats. Several regional atlases were produced and the second national breeding bird atlas was published in 2008 now covering the entire territory (including the Atlantic islands). In 2010 *Aves de Portugal – Ornitologia do Território Continental* was published. This landmark book was the first monograph on the birds of mainland Portugal in nearly 80 years.

As to the islands, early accounts about the birds of the Atlantic archipelagos of Azores and Madeira were made by the first settlers there in the 15th and 16th centuries. In those days, birds, and especially seabirds, were an essential resource as they provided meat, eggs, feathers and oil. Concerning this, the written chronicles of Fray Gaspar Frutuoso are very insightful and give a general idea of the size of the populations of some species and of the extent of the harvesting.

The first modern reports about the avifauna of these islands however were produced in the late 19th and early 20th centuries. As happened in mainland Portugal, most of the information, with one or two exceptions, was collected by foreign residents and visitors.

In mid-1960s the famous naturalist and ornithologist David Armitage Bannerman and his wife W. Mary Bannerman after visiting both archipelagos and also Cape Verde, Canaries and Selvagens, produced the first complete monographs about the birds of the Macaronesian islands. From the 1980s onward, two state bodies, the Parque Natural da Madeira and the university of the Azores, became the leading institutions in ornithological work in each archipelago. Other entities like Parque Ecológico do Funchal in Madeira have also been active. From the late 1990s onward, SPEA established offices in both archipelagos and has launched big conservation projects in the Azores, namely concerning the rare Azores Bullfinch, and started to work in close cooperation with Parque Natural da Madeira.

Today, as a result of the combined efforts of all the individuals involved, knowledge and awareness of Portuguese birds has increased immeasurably.

PRE-TOUR INFORMATION

Travelling to Portugal and travelling within Portugal is relatively easy. European Union nationals are allowed to enter the country bearing just a national identity card; people from other countries require a passport and, in some cases, a visa. Visitors are usually allowed a three month stay (this restriction does not apply to European countries that have signed the Schengen agreement), although a few nationals have more limited visiting privileges. Ask for details in your travel agency.

If travelling by car, you will need the following documents: personal ID, valid driving licence, car ownership papers (vehicle registration or equivalent and vehicle logbook or equivalent) and motor insurance certificate. It is advisable to inform your car insurance company, as they can arrange suitable cover if you wish insurance over and above the basic 'green card'. You may not be asked to stop at the border when entering from Spain, as customs posts have been withdrawn according to the rules of the Schengen Agreement, with people and goods being subject to free circulation.

In Lisbon, one can find the embassies or consulates of most countries. Furthermore, in Porto, there are consulates for several different countries. In the Algarve there are also some consular offices not only in Faro but also in other towns like Portimão, Vila Real de Santo António and Olhão. Consulates are also available in the archipelagos of Madeira and Azores at Funchal and at Ponta Delgada (São Miguel island), respectively. The addresses of all these offices are available in telephone books or can be found on the Internet (in Portuguese, embassy = *embaixada*, consular office = *consulado*) – see the 'Useful Contacts and Websites' section at the end of the book.

Portugal's currency is the Euro (€). If you are coming from outside the so-called Eurozone, it is advisable that you change some money before travelling. However, you can also do it upon arrival without major problems if you hold any major tradable currency.

No special vaccinations certificates are required from EU citizens wishing to enter Portugal. Citizens of some non-EU countries do have to show certificates of inoculation against certain diseases. If you think this might apply to you please contact your nearest Portuguese embassy or your travel agent for advice.

TRAVEL INFORMATION

Travelling to Portugal

There are three main ways of getting to Portugal: by air, by rail or by road. No car ferry or passenger lines regularly visit Portugal, but a few cruise ships call at Lisbon and Madeira and the car ferry from Plymouth to Santander and Bilbao in Spain brings car owners within a day's drive of mainland Portugal.

By air, there are regular flights from most of the larger European cities to Lisbon, Porto and Faro. There are also many cheap charter flights from European capitals to Portugal and your local travel agency can often arrange flight and car hire simultaneously. All the above mentioned airports are a bus ride away from their corresponding city centres; taxis are still relatively inexpensive.

Travelling by air to the archipelagos of Madeira and Azores is the usual way of getting to these offshore island groups. In the case of Madeira, there are daily flights from Lisbon and Porto and regular flights from many European cities. Flights from central Europe to the Azores usually land at Ponta Delgada, São Miguel island, but those leaving from Lisbon also include direct flights to the islands of Faial, Terceira and Santa Maria. The remaining islands can only be reached using local inter-island flights. Additionally, there are daily flights linking Madeira to the Azores. Please see the 'Useful Contacts and Websites' section at the end of the book for contact details.

By rail, there are daily trains from Paris and Madrid to Lisbon. These are 'normal' express overnight trains. There are plans to build a high-speed line from Madrid to Lisbon but as of 2013 its construction had not yet begun. The InterRail ticket and other special rail offers allow unlimited travel within Europe, including mainland Portugal.

Entering mainland Portugal by car normally means crossing Spain on the E80/A62, which leads from the French border through Valladolid and Salamanca (Spain) and enters mainland Portugal at Vilar Formoso. Two alternatives also exist. The E90/A5, which runs further south, from Madrid through Talavera and Mérida, crosses the border at Badajoz, near Elvas and the E1/A49, which comes from Seville and crosses the frontier over the bridge at Castro Marim, in the south-eastern corner of mainland Portugal. All these roads adhere to international standards and nowadays it is possible to drive from France to mainland Portugal using motorways only (it takes about 6 hours to drive through Spain from Irun at the French border to Vilar Formoso).

Travelling within Portugal

Certainly the most convenient way to travel in Portugal is by car; the road network is quite extensive and provides access to most areas, though in some of them it may be necessary to use rough tracks to approach the more remote and interesting sites. A good map (see next section) will allow you to travel throughout most of the country, with minimal navigation skills required.

In Portugal, the following categories of roads exist:

- Auto-estradas (motorways), usually known as A1, A2, etc. Relative to the size of the country they are numerous and provide easy and speedy access to most regions in the country. A toll must be paid when using them (about 8 Eurocents/km in 2013). It should be noted that on certain motorways, a special device is needed in order to be able to travel on them (see the 'Useful Contacts and Websites' section at the end of the book for details about tolls).
- Main Roads (usually marked IP or IC on sign posts). These are modern roads, toll-free, making for rapid connections between the main cities.
- National Roads (N1, N123, N375, etc.), which form a dense network all over the country. Some of these roads are now called 'regional' and are marked with R instead of N.
- Municipal Roads, of variable quality and in most cases unnumbered, sometimes marked with M.

For those using right-hand drive cars it is advisable to cover the prism of the headlights with a triangle of dark adhesive tape. Never stop on the left side of the road as the natural tendency is to set off again on that side. This can lead to situations ranging from the comic to the tragic. Cars on roundabouts have right of way unless clearly stated otherwise.

In case you need to do so, hiring a car should not be a problem except perhaps in the Azores, where it is advisable to make a reservation in advance (especially in the smaller islands), and also during the summer months at the more tourist-saturated places like the Algarve or Madeira. Cars can be rented for short or long periods and can often be arranged through one's own travel agency. Most international companies are represented in the country. Besides there are also a lot of local companies that can sometimes offer better prices. Browse the net before deciding.

It is worth pointing out that road deaths are alarmingly high in Portugal and great care should be taken using roundabouts or minor roads, or when parking to watch interesting species. The maximum speed limits for passenger cars are the following: 120kph on motorways, 90kph on most other roads and 50kph in urban zones (including small villages).

For local movement within cities, taxis are a useful form of transport. They are always equipped with digital taximeters.

Trains

Travelling by rail can be an interesting option but is only available in mainland Portugal as there are no trains on the islands.

Tickets are cheap, compared with most west European countries. If not travelling with your own car, it is sensible to travel by train and then rent a car in the area to be explored, as petrol and motorway tolls can prove very expensive. It should be emphasised, however, that the rail network in mainland Portugal is not very extensive. Additionally, service has been withdrawn on a number of lines, especially in the eastern half of the country, thus making railway an even less convenient mode of transportation; the only important routes run from Lisbon to Porto and from Lisbon to the Algarve, and these are of little use to the birdwatcher.

For most sites, rail travel is not a viable option. Nevertheless, some of

them can be reached by train, particularly Minho estuary (trains running from Porto to Valença), Portas de Ródão (served by intercity trains linking Lisbon to Covilhã), Paul do Boquilobo (served by local trains on the Lisbon-Porto line), the beaches on Costa do Estoril (served by suburban trains from Lisbon to Cascais), Corroios on Tagus estuary (served by local trains from Lisbon and Setúbal) as well as Faro, Ria de Alvor, Ria Formosa and Castro Marim (on the Algarve coastal line). All these sites can easily be explored on foot or using local transport from the nearest railway stations. Other towns, like Évora and Castelo Branco (close to International Tagus) can also be reached by train, but the enormous sites nearby are probably very difficult to explore completely on foot. The schedules can and should be checked against the latest wall timetables at the station before departure; these can also be found in the Internet (see the 'Useful Contacts and Websites' section at the end of the book).
The main train types are:

- Alfa Pendular – quality express trains connecting Lisbon to Porto, Braga and the Algarve
- Intercidades (Intercity) – fast trains running from Lisbon and Porto to several major cities
- Regional – slow trains which call at most stations
- Suburbano – suburban trains running in the Lisbon, Porto and Coimbra areas.

Buses

Buses can also be used to reach several interesting areas, although after arrival extensive walking may be required to reach the more remote sites, where transport facilities may be non-existent. The most important express bus network is run by 'Rede Expressos'. General information and tickets can be obtained at their website (see address at the end of the book).

Bus transportation on the islands is limited and information should be sought locally.

Boats

Boat travel is not very important on the mainland, except to get to islands which are not served by any other transport, such as the Berlengas or the barrier islands in Ria Formosa. Boats can also be used to cross the river Tagus just in front of Lisbon and the river Sado at Setúbal.

It is on the archipelagos of Madeira and Azores that boats emerge as a relevant form of transport. In the case of Madeira, taking the daily boat to Porto Santo is probably the most convenient way to get to this smaller island. It is also the only way of getting to the Desertas and the remote Selvagens islands. In the Azores, inter-island boat crossings link the islands to each other, especially among the islands of the so-called central group. Note however that these crossings are subject to weather conditions. Please refer to the 'Useful Contacts and Websites' section at the end of the book for more details about boat companies and timetables.

STAYING IN PORTUGAL

Accommodation

Accommodation is widely available in Portugal and a range of hotels exists to suit all pockets.

Large hotels are more numerous in the main cities. Small hotels ('estalagem', 'albergaria', 'residencial', 'pensão') are also available in many towns. In mainland Portugal, the coastal areas are the most important tourist attractions and there is a plethora of possibilities, from camping sites and pensions to five star hotels and apartment hotels. However, in July and August many hotels may be fully booked. This contrasts with what happens in some inland areas where accommodation may be more limited. In the more remote Azorean islands there are fewer choices so it is best to check the possibilities before going and to book in advance where possible.

If staying at smaller more modest hotels, it is advisable to 'shop around' before deciding on a particular hotel. The tourist office in every town will happily help you with maps, brochures and lists of seasonal events.

It is also possible to book accommodation online. Many hotels, even small ones, accept online reservations and by booking in advance one saves time and sometimes money.

Money

The bank network is extensive and banks and cash machines facilities exist in most towns and villages. Banks are open on weekdays, from 08.30 to 15.00 and will readily accept US Dollars and most west European currencies. Exchange offices also exist at the airports and are open until later in the day. Many stores, hotels, pensions, restaurants and petrol stations accept VISA and most other common credit cards, as well as Eurocheques, though some restaurants and shops do not accept cards at all. It is therefore recommended that you carry some paper money with you.

Food

From the gastronomical point of view, Portugal is extremely interesting. Most regions have their own typical dishes and while it is not possible to describe them in detail here, sensible gourmands are recommended to experiment with Portuguese food. It is advisable to purchase a small gastronomical pocket guide. A 'dose' (portion) of any typical dish is likely to shock you by its size and quantity and in many places you can request a 'meia-dose' (half-portion) which many people find quite sufficient – in these cases a full portion will be enough for two. A 'touristic menu' is available at most restaurants and consists of a full meal at a fixed price.

Restaurants are to be found throughout the country and most localities have their own typical restaurant, where tourists are particularly welcome, typical dishes are served and waiters are extremely helpful. Although the opening times vary, meals are usually served from 12.00 to 15.00 and from 19.00 to 23.00. International fast food restaurants are widespread, being found in major cities and in the Algarve. Besides, there are many restaurants in shopping centres that are open continuously throughout the day until 22.00 or 23.00.

Roadside 'cafés' are probably more widespread in Portugal than anywhere else in Europe and are open from early morning to 20.00 or 22.00 or even midnight. Here you may eat a sandwich or a small cake and drink juice, a glass of milk or just a coffee at any time of the day; there are often chairs available to sit outside. These 'cafés' may be found even in tiny villages. Spend some time resting at these oases – people can be even more interesting than birds!

Finally, there are small to large supermarkets all over the country, where other types of food may be purchased, like bread, cans of food and large bottles of juice and water. Like most stores, they are usually open from 09.00 to 19.00. However larger supermarkets may be open until later (typically until 22.00 or even later for those located in big shopping centres; the latter are also open during weekends).

Wine

Portugal's wines are famed far beyond her frontiers and should not be ignored by anyone visiting the country. Apart from the renowned Port and Madeira, there are numerous excellent wines from all parts of Portugal available. In the northern part of the country, 'vinho verde' is a unique treat and should be tasted there as it does not 'travel' well. In the centre, the best wines are certainly the ones from Dão and Bairrada regions, while in the more southerly and drier Alentejo, some of the best ones come from Borba, Redondo and Reguengos de Monsaraz.

However, remember that alcohol is still one of the main causes of road accidents in Portugal, so the rule is, if you drink simply do not drive. The police rightly consider drunken driving as a major offence. It is also worth remembering that summer heat plus 'one too many' could end your birding, forever. The blood alcohol limit is 0.5 g/l.

Safety and theft

In terms of personal safety, Portugal is widely regarded as less dangerous than many other southern European countries and, until recently, Lisbon was regarded as the safest European capital. Unfortunately, as is the case in so many other cities today, crime has increased markedly in the last few years, above all drug related crime. It is therefore necessary to be sensible, especially in the whole area within 50km of Lisbon, in most large cities and in the Algarve. Do not leave valuables in your car. If you must do so, conceal them well before parking. Car radios are stolen very frequently and the same precautions should be kept in mind. Avoid carrying too much money at once or in the same pocket.

In the more remote countryside, crime is almost non-existent. However, care should always be taken. In those places you may be asked what you are doing. Common courtesy, mime and a bird book will usually suffice to persuade the enquirer that you are merely another mad 'forasteiro' (someone not from the local area).

Cattle

This warning is not to be taken lightly.

Cattle form an important component of agricultural activities in mainland Portugal. Bull fights are almost as popular here as they are in Spain and these huge black fighting beasts wander freely in the many areas of farmland, which, theoretically, are fully fenced. Extreme care is advisable. When driving, you may occasionally find the track blocked by wire gates. These gates are normally used to avoid cattle moving from one field into the next and if not locked, there should be no problem in opening them and driving through. However, please close them immediately, as the bulls may escape and on return you might have to face an angry owner or an even angrier bull. Not all bulls may be recognisable as such, so be extra vigilant. When driving in areas with cattle (particularly in the Tagus estuary), avoid jumping over fences and passing close to the bulls.

It is also common to find the road blocked by a flock of sheep crossing or simply walking along the road, even in the middle of villages. Be patient and take care, even if the animals are occupying the whole width of the road; those few minutes of pastoral bliss may bring some high flying raptor into view.

Extra care should be taken while driving in the Azores as it is rather frequent to find cattle in the middle of the road, looming suddenly out of the mist and rain. The same applies to the highlands of northern mainland Portugal, especially those of serra do Gerês, where sightings of cattle wandering freely along the roads are not unusual.

Hunting

Hunting activities in Portugal reflect an ancient tradition once found throughout Europe and are an established pastime in Portuguese society. Although officially there are 300,000 registered hunters, the number of active ones is probably half of that.

The hunting calendar changes according to the species. In general terms, it is permitted to hunt from mid-August to late February. According to the law, hunting may only occur on Thursdays, Sundays and holidays. Nevertheless, private hunting areas may have a different schedule.

For many years, following the 1974 revolution, hunting was a free-for-all event. It was the so-called 'free regime', since nobody had to pay for hunting at a specific site. However, in recent years more and more private hunting areas have been created and the operating rules are determined by the 'Hunting Law', which aims at bringing some order to a previously anarchic activity. Unfortunately, this 'regime' has not been well received by many traditional hunters, who are now either asked for a fee or forbidden to hunt in those areas that, due to agricultural decline, have been quickly turned into private hunting areas. Furthermore, so-called 'predator control' is a common practice in some areas, where many birds of prey and mammals are still illegally shot, in the name of 'game conservation'.

If you wish to visit a hunting area (usually fenced and marked with red-and-white metal plates), it is advisable to contact the owner or the game keepers before entering or at least before beginning birdwatching.

Private land

Not very long ago it was easy to walk and drive in the field. For both good and bad reasons, it is now basically very hard to do so. A good number of private properties are now fenced and access is blocked by gates that are usually kept locked. This has happened mainly in the southern half of mainland Portugal and is particularly obvious around the two major estuaries of Tagus and Sado and over most of the Alentejo province. It is not advisable to walk through these properties without having first obtained permission. In some places owners or workers may get angry and tell you to go away in no uncertain manner. On the other hand, if you seek permission in advance the odds are that you will be allowed free passage.

Fire

Mainland Portugal has a Mediterranean climate and fire is a natural agent and hazard. During the 20th century, Portuguese forests have been subject to marked changes; the original cover of oaks has been largely destroyed and partly replaced by pine and eucalyptus plantations. However, the combination of the climate, poor forestry management and land abandonment have turned these woods into powder kegs, particularly in summer.

In the last few years, enormous forests have been destroyed by fire and though there is no doubt that natural conditions have changed markedly and predispose these forests to fire, it has been shown that a good number of fires start by negligence or are set by arsonists. Fire has almost turned into a 'tradition' during the summer months (June through September) and the consequences have been dramatic. Over the 10 year period (2001-2010) an average of 147,000ha of woodland and scrub was burned every year (over 1% of mainland Portugal's surface) with a record of over 425,000ha being decimated in 2003.

Fire is also a problem on Madeira island, where in the last few years large areas have burned including the breeding grounds of the rare and endemic Zino's Petrel.

Apart from the dramatic ecological consequences, large fires can block some important roads, lead to delays in your journey or can be life-threatening. The northern and central parts of Portugal are more wooded and, as such, are the most sensitive areas. Care should be taken never to light picnic fires, especially in pinewoods.

WEATHER AND CLOTHING

Mainland Portugal lies on the periphery of south-west Europe, with a long coastline and a climate dominated by the Azores subtropical anticyclone. Because it stretches through five degrees of latitude, weather varies markedly from north to south, with a mild Atlantic regime obvious in the north-west and a distinctive Mediterranean type of climate elsewhere. The interior of Portugal experiences hot summers and cool winters and is characterised by a continental climatic regime, a feature compounded by the proximity of Spain's high meseta.

The weather varies greatly according to season and altitude. In the south the climate is basically Mediterranean, which calls for light clothes with a pullover for cooler evenings. Obviously early morning seawatches and walks through thorny garrigue need warmer and more durable clothing respectively. Winters in the south may be wet and chilly and rainproof gear may be needed. Surprisingly perhaps, the Alentejo, known more for its hot dry summers, can suffer long periods of very cold weather. Pack clothing to suit these conditions.

In the north-west, the climate basically resembles that of southern Britain; equally British, the weather in this part of mainland Portugal is notoriously unpredictable and clothing should be chosen accordingly. In the north-east, around Bragança, summers may be unbearably hot and winter temperatures may drop well below zero when it is essential to wear warm and waterproof clothing. Remember that this area is like any other montane region – subject to sudden changes in weather. Be careful to check weather forecasts before setting out to walk across the mountain tops.

Madeira is an island of contrasts: on the northern slopes, including the tops, the climate is wet, fog is frequent and it often rains. The southern slopes, namely around Funchal and also in the far eastern Ponta de São Lourenço, enjoy a climate which is drier and sunnier. If intending to venture into the mountains or along the northern slopes, please take these aspects of weather vagaries into account and inform your hotel of where you are going. Porto Santo has a dry climate, much like southern Madeira.

The Azores have a generally wet and mild climate, and although temperatures rarely drop to freezing point, the number of sunny days is much lower than on the mainland. There are, however, some contrasts between islands, with Santa Maria and Graciosa, the two least-mountainous islands, being the driest and with Flores and Corvo, in the remote north-west, being the wettest. In any case, if travelling to the Azores, pack for adverse conditions. Note that the archipelago is often hit by severe storms and even hurricanes.

HEALTH AND MEDICAL FACILITIES

As is the case throughout most of Europe, the chances of contracting a serious disease in Portugal are remote.

It is not necessary to be vaccinated against any major contagious diseases and the worst problem may be mild diarrhoea and sunburn. On the other hand, nationals other than those of the EU may require valid vaccination documents to enter Portugal.

Almost everywhere tap water is perfectly safe to drink but in some places it may be distasteful, especially in the south. Bottled water is an excellent and cheap alternative. Like everywhere that is warm in summer, there is a slightly heightened risk of eating food that may cause gastro-enteritis. The treatment, if you do get stomach upset or diarrhoea, is available in all pharmacies. Obviously persistent symptoms should not be ignored.

Sunburn (and even sunstroke) should be taken seriously, not only because it may spoil your trip but because serious burns can have dangerous consequences. Always wear a hat even in winter, use strong barrier creams especially if you are light-skinned and avoid birdwatching between 13.00 and 17.00 in summer.

Problems with snakebites are very, very rare but obviously, if bitten, it is essential to obtain medical assistance. Scorpions do exist but the risk of being stung is minimal. A mosquito repellent is useful in summer especially if you visit wetland areas. Some cases of dengue fever were reported from Madeira in 2012 but this is a new issue and its extent is presently unknown. To be on the safe side, extra care with mosquitos should be taken while visiting the island.

There are hospitals all over the country but as elsewhere the most comprehensive medical care is available in the larger cities, notably in Lisbon, Porto and Coimbra. It is usually not difficult to find a doctor and pharmacies are widespread. If a pharmacy is needed during the night, enquire which pharmacy is on duty on that day – most hotels have this information, which can also be checked at any pharmacy (a calendar is usually placed at the window, so that it can be viewed at any time) or on the Internet – please see the 'Useful Contacts and Websites' section at the end of the book.

EU citizens should carry the European Health Insurance Card (EHIC) from their home country. It entitles them to receive medical treatment for free or at a reduced cost. It is also advisable to take out some form of medical insurance even for short periods. Ensure that your travel insurance covers both medical costs and repatriation in case of need.

If an emergency arises, 112 is the national emergency telephone number.

MAPS

To get the best out of this book, a general map of the country should be purchased to identify the different roads mentioned and to locate the principal towns and cities. There are many large-scale maps (1:300,000 or 1:400,000) available, though the following ones are recommended:

- Automóvel Club de Portugal (ACP), scale 1:350,000, available in bookshops or directly from ACP (see the 'Useful Contacts and Websites' section at the end of the book).
- Michelin map no. 733 (Portugal and Madeira) – scale 1:400,000.

For many sites, however, a more detailed map is invaluable to reach specific places. It is possible to purchase small-scale maps for any part of the country and the following scales are available: 1:100,000; 1:50,000 and 1:25,000.

Maps of the first two scales can be purchased at Instituto Geográfico Português (IGP), Rua Artilharia Um, 107, in Lisbon (near Amoreiras). Ask for the 'Venda de Cartas' (map sales section). It is open from 09.00 to 16.00 and the maps, as of 2013, cost about €2.55 each (you may be asked for identification at entrance). Information about these maps can also be obtained on the internet (see the 'Useful Contacts and Websites' section at the end of the book).

The 1:25,000 maps are a military edition and can be obtained from Instituto Geográfico do Exército (IGEOE), Avenida Dr. Alfredo Bensaúde, in Lisbon (not far from Oriente railway station). To get there, leave Lisbon city centre following signs to the airport. On reaching the large roundabout which is just 500m before the airport, follow the signs to A1 and Porto and proceed for about 1.5km, turning right after the GALP petrol station. Continuing along a wide avenue for about 500m, one reaches a large building on the left, called Instituto Geográfico do Exército. Enter and ask for the 'Venda de Cartas' (Map Sales Section). It is open from 09.00 to 17.00 (16.00 on Fridays) at the unit price, as of 2013, of €6.40 (identification is necessary to enter the building). Like in the previous case, information about these maps can also be obtained on the internet (see the 'Useful Contacts and Websites' section at the end of the book).

To facilitate identification of the correct small scale map to be purchased for each of the main sites in this book, the corresponding number is given at the beginning of each site description.

Maps for Madeira and Azores are available at both institutes as well.

GEOGRAPHY AND VEGETATION

Mainland Portugal is a quadrilateral of land stretching from 42° 09′ N to 36° 58′ N on the western coast of Iberia and occupying an area of 89,015km². In terms of physical geography, a little under half of the land lies below 200m and there are several distinct regions: the montane areas of the north, the eastern plains, the basins of the Tagus, Sado and Guadiana rivers and the peripheral western and southern borders. These areas, each with its own distinct avifauna, have been moulded by three factors: by a seismically active underlying structure, by rock type and by climate. Leaving aside the tectonic aspects for which Portugal is famous (or infamous, depending upon how you view the great Lisbon earthquake of 1755), there are three rock types that fundamentally affect the landscape and vegetation: calcareous, granitic and schistose.

The calcareous limestone areas have the distinct and beautiful morphology of karst, well seen in the Estremadura ranges of west-central Portugal (e.g. serra da Arrábida). Relief underlain by granite forms the rather spectacular scenery of serras do Gerês, Estrela, Sintra and part of the eastern Alentejo. The gradients are steep, giving deep valleys, cascades and waterfalls. Rivers crossing it tend to produce characteristic 'dog-leg' turns as they follow the fracture lines of the granite. The classic steppe landscape of Baixo Alentejo with its rolling rounded hillocks, long vistas and fields of Little Bustards and larks is mostly underlain by schistose rock. The few rivers that wander across this peneplain do so in long wide, shallow valleys. In some Beira Baixa areas, especially close to the Spanish border, the schist produces a more contrasting landscape with deep cut valleys and high river banks ideal for cliff-nesting birds.

Past climates both cold and hot have left tangible marks upon the landscape of Portugal from the terra rosa of bauxite deposits to the glaciated U-shaped valley of the river Zêzere in the serra da Estrela. This valley is one of the best examples of its kind in the world.

Mainland Portugal's coastline, an ancient abraded continental platform, is marked by its regular outline. The long sweeps of sand and dune are broken, occasionally spectacularly, by capes such as Cabo da Roca and estuaries such as the Tagus. Sandbanks and spits are common and, together with slow uplift have produced lagoons, rias and marshland such as Ria de Aveiro, Ria Formosa, Lagoa de Óbidos and Lagoa de Santo André.

The vegetation of mainland Portugal, just before man's intervention, was dominated by oak species, deciduous in the cooler wetter north, evergreen in the warmer drier south. This pattern still persists, though much modified by man and local mesoclimatic effects. Portuguese oak woods are characteristically complex and diversified with a wide variation in avifauna. In the north-west the dominant tree is Pedunculate Oak (*Quercus robur*), while in the higher reaches of the north-east, Pyrenean Oak (*Q. pyrenaica*) is more common. In the south and along the International Douro, the oak woods consist largely of evergreen oaks like Holm Oak (*Q. rotundifolia*). In the west-central fringes Portuguese Oak (*Q. faginea*) can be found and south of the Tagus Cork Oak (*Q. suber*) and Holm Oak are increasingly the only two large oaks to be found. Although still abundant in the south, many Holm and Cork Oak woods have however disappeared under a swathe of wheat in Beira Baixa, Alentejo and the Algarve hills – these two species are now protected by law and may not be cut without permission.

Loss of the understorey vegetation to secondary formations has led to environmental degradation, while vast plantations of eucalyptus trees (*Eucalyptus globulus* and other species) in areas of fragile, friable soil have eliminated many natural springs, removed top soil and reduced biodiversity. Fire and grazing have also taken their toll with oak groves successively reduced to thickets of maquis with Tree Heather (*Erica arborea*), pistachio (*Pistacia sp.*) and Strawberry Tree (*Arbutus unedo*). Further loss of biodiversity has led to the spread of dense areas of garrigue – dominated by the small Kermes Oak (*Quercus coccifera*) and to hillsides swathed in monotonous swards of thorny scrub plants like gorse (*Ulex sp.*), heather (*Erica sp.*) and sun rose (*Cistus sp.*).

Much of the native flora of the Algarve has been replaced by orchard species. Many fruit trees typical of Mediterranean landscapes were brought by the Phoenicians and Arabs a long time ago. Olive (*Olea europaea*), carob (*Ceratonia sp.*), Almond (*Prunus dulcis*) and Fig (*Ficus carica*) still abound and offer excellent habitat for both breeding and migrant birds.

Maritime Pine (*Pinus pinaster*) has been sporadically planted in Portugal since the 13th century. Initially this was carried out in the coastal areas of central Portugal to prevent expansion of the coastal sand dunes and to protect agriculture land. In more recent times, beginning around the middle decades of the 19th century, widespread planting of Maritime Pine has displaced the old Pyrenean Oak and Portuguese Oak forests. These plantations were established with economic profit in mind, often against the wishes of the local population: one hundred and fifty years on, these pine forests now form some of the most extensive pinewood coverage in Europe. Umbrella Pines (*Pinus pinea*) have also been widely planted, mainly in the Sado basin.

Information about the geography and vegetation of the Azores and Madeira can be found in the corresponding sections.

WHEN TO GO

Mainland Portugal

From a birdwatcher's point of view, mainland Portugal is interesting at all times of the year. Nevertheless, each season has its own specialities.

Spring

For breeding and migrating birds, March to June is generally the best and most pleasant season to visit mainland Portugal. Spring starts early and by March nearly all residents are in full breeding activity and by mid-April most summer migrants have arrived. A few species like White-rumped Swift, Rufous Bush Chat, Melodious Warbler and Red-backed Shrike typically arrive later and are not easily seen before May. Because Portugal stretches through several degrees of latitude, there is about a one month delay between the seasons, south and north. This means that in the south the best months for breeding birds are probably April and May, with bird activity decreasing markedly in June. Regardless of birds, June in any case can be unbearably hot away from the coast, south of Lisbon. But travel just a few hundred kilometres north and June is a superb month to be in the field, with breeding activity at its height.

Passage migrants can also be found throughout the spring, although numbers vary markedly from year to year, depending on weather conditions. Often this means that bird diversity can reach its highest in early spring, as winter visitors mix with summer visitors and passage migrants. Northward passage typically peaks in April and early May and can include vast flocks of Black-tailed Godwits, Grey Plovers and Knot flying over estuarine areas or north along the coast. Common and Black Terns pass north from April onwards and occasionally heavy passage is still obvious in May. Falls of passerines on migration occur at times, especially after prolonged winds from the east. However, it is clear that the event called spring migration in Iberia is intense, rapid and well-timed, usually taking place a little to the east of Portugal.

In June most coastal wetlands are quiet and the number of birds present is at its lowest. The estuaries are almost empty, and most birds are concentrated in the surrounding saltpans, rice fields and reedbeds, where breeding takes place.

Summer

During July there are good birdwatching opportunities in highlands areas, where temperatures are milder and the breeding season is typically delayed, relative to the lowlands. A visit to serras do Gerês, Montesinho or Estrela can prove rewarding for finding birds that are typical at higher altitude, like Rock Thrush and Ortolan Bunting; they may still be singing. Elsewhere this season can be unpleasantly hot inland and overcrowded at some coastal locations (particularly close to beaches). Most songbirds are silent and often difficult to locate. However, many coastal wetlands attract a number of waterbirds, namely egrets, ducks, gulls, terns and some waders. Many of these birds arrive following dispersion movements from their breeding sites. Some are simply non-breeders remaining around during the breeding season or adults and non-breeding birds that have returned earlier from their northern locations; some of the former may still display part of their colourful breeding plumage.

By August, however, post-breeding migration starts to be noticeable and it becomes more intense during the second half of the month. Of those species that do not breed in mainland Portugal or do so only in tiny numbers, waders are probably the most noteworthy. As summer advances, the number of waterbirds starts to increase on most coastal wetlands, as the adults are now followed by juveniles, making these locations particularly interesting. As to small birds, movements of trans-Saharan migrants reaches its peak in September, when large numbers can be found at most coastal locations, particularly in the Algarve, with flycatchers and warblers especially abundant. From mid-September, though somewhat dependent on weather, the movement of soaring birds at Cabo de São Vicente area becomes intense.

Autumn

As October sets in, post-breeding migration is in full force – an annual and evident phenomenon in Portugal, especially along the coast. At this time, many trans-Saharan migrants can still be found. Most passage migrants are now adults in winter plumage or immatures in first-winter plumage and can provide a real identification challenge. Storm-driven seabirds may pass south in spectacular numbers. Coastal wetlands are now full of gulls, waders and other waterbirds. Raptors can be seen circling southwards at some headlands, especially at Cabo de São Vicente, where on certain days hundreds of birds turn up. Sightings of migrants often depend on weather conditions, but in general their number in autumn is far greater than in spring and 'falls' of migrants may occur when winds from the north-east prevail for a few days. Autumn is also the time of year when most rarities from Siberia or North America turn up. Birds are still on the move in mid-November, by which time the weather may be wet and windy. Late November marks the end of most active migration but by then most wintering birds have already arrived at their winter quarters. The estuaries are now full of birds, with numbers typically continuing to increase as mid-winter approaches.

Winter

Winter can be wet but is usually mild and can produce interesting and unexpected species, especially waterbirds. At this time of year, birds tend to concentrate in the southern plains and on the coastal wetlands and it is not impossible to see 80 or even 100 species in one day at certain places. Enormous flocks of gulls and waders fill the estuaries and in some of them there are parties of waterfowl as well. Many beaches are also used as overnight roosts by gulls and often attract unusual species. Inland the number of birds is more variable but in general it can be stated that the best places to watch are in the south – parties of Cranes can be seen at a few places in the Alentejo province, where it is also possible to find large flocks of lapwings, larks, pipits, thrushes and finches.

There are many reservoirs used for irrigation, some of which can hold considerable numbers of wintering waterfowl. Some large soaring birds of prey start their courtship display as early as January and vocal activity of Eagle Owl is at its maximum from mid-December until

January. Some resident passerines can be heard in full song during January. In the highlands, on the contrary, there are very few birds at this time of year, but a few winter specialities, like Alpine Accentor, Wallcreeper and Snow Bunting sometimes appear on remote hilltops. In the south, a few migrants typically start to arrive in February, notably hirundines, Lesser Kestrel and Great Spotted Cuckoo and by mid-March, Great Bustards males become active at their 'leks', by which time most resident songbirds may be already in full breeding activity.

Azores and Madeira

The calendar in the archipelagos of Azores and Madeira is somewhat different from that of Mainland Portugal. Most landbirds found there are resident and can be seen year round. Many seabirds however, are clearly summer visitors and are best looked for in spring and summer, although some interesting species are winter breeders. The arrival of European wintering birds is low-key, both in Madeira and the Azores, but some migration may be noticed. For those interested in rarities, especially Nearctic rarities, the best time to visit the Azores is the post-breeding period from September to November. More detailed information can be found in the sections dedicated to the islands.

Seabirds and seawatching

Mainland Portugal's coastline is 800km long from the mouth of the river Minho in the north-west to the mouth of the Guadiana in the south-east. From a geographical point of view, it appears eminently suitable for seawatching (see map), however its potential is much reduced by meteorological patterns.

In a typical winter the weather is dominated by an anticyclone lying north of the Azores, bringing dry cool north-east winds. Seabirds are blown clear of the coast, though wintering Sandwich Terns and odd Great and Arctic Skuas, as well as Mediterranean Gulls and Razorbills, may be seen inshore. Offshore, Great and Arctic Skuas appear sometimes in large numbers flying north after wintering in Portuguese waters, though spring sightings of Pomarine and Long-tailed Skuas are rare. On average every fortnight this weather pattern is modified by low pressure at high altitude which brings onshore south-westerly winds, torrential rains and excellent seawatching: Leach's Petrel, Mediterranean Gull, Little Gull and Great and Arctic Skuas, appear, as well as rarer species. If the winds persist, Great Skuas may appear in three digit numbers.

Summer is characterised by the Azores anticyclone with pressure high from May to October; for four to six months, the *nortada* (a northerly airstream) may dominate with clear skies and gusty offshore winds. Seawatching may be consistently disappointing in early summer though usually Cory's and Balearic Shearwaters, as well as odd skuas may be seen from shore. However seawatching becomes more interesting as the season advances.

Both in spring and in autumn the weather is transitional-cyclonic, though in recent years spring has been drier and cooler. Frontal weather associated with Atlantic low pressure systems does occur and with onshore winds can produce very good seawatching conditions with (in autumn) hundreds of Storm Petrels, Sabine's Gull and skuas. However around latitude 40° N, frontal events are short-lived and observations need to be made within twelve hours of the passage of the front. In spring, similar weather may bring heavy passage of Gannets, Great and Arctic Skuas and terns, as well as the ubiquitous Cory's and Balearic Shearwaters. Seawatching from headlands in autumn may reveal heavy movements of seabirds, especially of Gannets and auks, well into December. In some years, Great Skuas may pass southwards in thousands, an awesome sight as party upon party move purposefully low over the sea within metres of the shore.

Despite some exciting seawatching from land in recent years, the predominance of the offshore winds in mainland Portugal as mentioned above means that seabirds are actually best observed from a boat. Fortunately, the self-same offshore winds are largely responsible for a remarkable upwelling of cold enriched water along the 200m bathymetric contour where the continental shelf slopes abruptly to 1000m. This bathymetric contour line runs relatively close to the Portuguese coast and can be easily reached by boat in about 3-5 hours at most. From July to September, Wilson's Petrels are very regular in small numbers and skuas, including Long-tailed Skua, have been observed closely. Sabine's Gull are seen annually and of course Cory's, Great and Sooty Shearwaters are guaranteed. Over a thousand Storm Petrels have been seen together and moulting Balearic Shearwaters are always present in July and August. In recent years, pelagic transects between mainland Portugal's and Macaronesian waters have revealed that Bulwer's Petrel and Macaronesian Shearwater are regular to within 90km of the Portuguese coast. White-faced Petrel occurs annually, especially in the region of the Gorringe bank, which lies about 250km south-west of Cabo de São Vicente. Besides this *embarass de richesse* there are records of Red-billed Tropicbird and petrels of the *Pterodroma* genus, either Zino's or Fea's, within 300km of the Portuguese coastline.

Off the coast of central mainland Portugal, the small Berlengas archipelago holds an interesting set of breeding seabirds including the only European colony of Madeiran Storm Petrel outside the Macaronesian islands. There are a few hundred pairs of Cory's Shearwater breeding there as well as large numbers of Yellow-legged Gulls. These islands are also the southernmost European limit for breeding Guillemot, but sadly, the local population is on the verge of extinction.

Seabird populations breeding in the Macaronesian archipelagos are of the utmost importance. In a worldwide context, they hold globally relevant populations of Cory's Shearwater, while in a European context the breeding populations of White-faced and Bulwer's Petrels are noteworthy. Passage of seabirds occurs off both archipelagos but wintering seabirds are not very noticeable in either archipelago. However, there are sometimes good numbers of Great Shearwaters in Azorean waters in late summer.

INTRODUCTION TO SITE INFORMATION

Geographically this book covers the whole of Portugal including the Atlantic archipelagos of Azores and Madeira.

The mainland sites have been grouped into chapters corresponding approximately to the five main regions into which the territory is usually divided (see map):

- North, which covers the area north of the river Douro; the main city in the region is Porto;
- Centre, corresponding broadly to the area between the rivers Douro and Tagus; the university town of Coimbra is the largest in the region;
- Lisbon region, which covers the country's capital and the sites lying within about 40km of the city;
- Alentejo, which extends from the river Tagus south to the Caldeirão and Monchique hills, with the historic city of Évora lying just in the middle;
- Algarve, which forms a rectangle along Portugal's south coast; its main town is Faro.

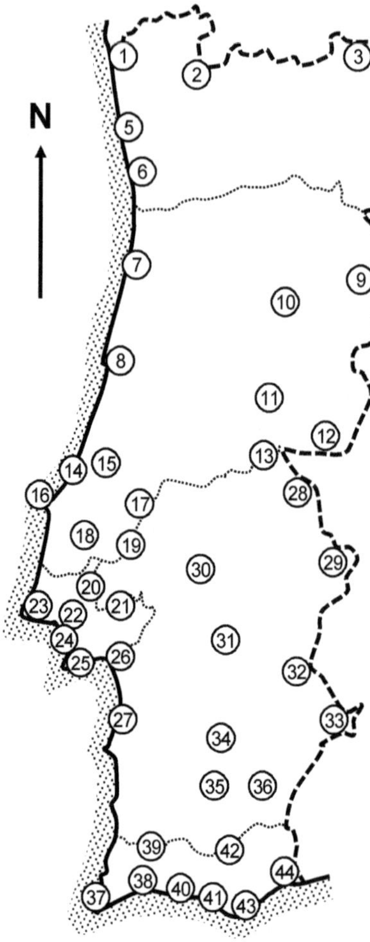

1. Minho Estuary
2. Serra do Gerês
3. Serra de Montesinho
4. International Douro
5. Cávado Estuary
6. Porto and Surroundings
7. Ria de Aveiro
8. Lower Mondego
9. Ribacoa
10. Serra da Estrela
11. Serra da Gardunha
12. International Tagus
13. Portas de Ródão
14. Lagoa de Óbidos
15. Paul de Tornada
16. Peniche and Berlengas
17. Paul do Boquilobo
18. Serra de Montejunto
19. Escaroupim
20. Tagus Estuary
21. Paul da Barroca
22. Lisbon
23. Costa do Estoril and Sintra
24. Lagoa de Albufeira
25. Cabo Espichel and Arrábida
26. Sado Estuary
27. Lagoa de Santo André
28. Serra de São Mamede
29. Elvas Plains
30. Mora and Montargil
31. Évora
32. Mourão and Alqueva
33. Barrancos
34. Lagoa dos Patos
35. Castro Verde
36. Mértola
37. Sagres and Cabo São Vicente
38. Ria de Alvor
39. Serra de Monchique
40. Lagoa dos Salgados
41. Vilamoura and Vale do Lobo
42. Serra do Caldeirão
43. Ria Formosa
44. Castro Marim

A selection of the best birding sites is included in each chapter. A total of 44 sites are described for the mainland (see map). The main ones are described in greater detail and the strategy section has often been divided in two or more subsections in order to make exploration easier. For additional sites only a short description is provided.

The birding information for Azores and Madeira is presented island by island. The section for Madeira has been extended in this edition, in order to cover the uninhabited Desertas and Selvagens archipelagos (there are now 4 chapters for this region). The chapter on the Azores is totally new. It was the authors' decision to add it now that this archipelago has become a key site for European birders seeking to increase their Western Palearctic lists with Nearctic rarities. Nine sections have been provided for the Azores, one for each island – here also the amount of detail varies, depending on the size and bird diversity of each island.

The maps of the larger areas in this guide only show main roads and tracks. It is possible to buy some of the large scale maps as described in the Maps section or view details in web applications like Google Maps. The geographical coordinates are provided for all sites on the mainland, in order to make access easier for users with navigation devices – these coordinates refer to the main town or central location of each site, unless stated otherwise.

Some geographical terms like 'serra', for instance, have been retained in Portuguese as the English translation is not straightforward ('serra' can be translated as mountain but in Portuguese there is a distinction between mountain and serra, the latter being a chain of hills not surpassing 2000m). The same with 'lagoa' which on the mainland is used for coastal lagoons but which in the Azores and on serra da Estrela is the name given to inland lakes away from the coast. A list of the most common Portuguese landscape descriptors used in the site texts is provided as an appendix at the end of the book.

An annotated list of some of the more relevant species occurring in Portugal is provided at the end of the book as well as a complete list of all the birds seen in Portugal, including all of the Nearctic vagrants that have turned up in the Azores in recent years.

An appendix called 'Useful Contacts and Websites' is also provided to enable easy access to information either logistical or site specific.

NORTH

Minho Estuary

Geographical coordinates: 41°52′36.20″N, 8°50′13.00″W

Lying in the climatically wetter and cooler north-western corner of the country, forming the border with Galicia in Spain, this small estuary is one of the least studied coastal wetlands in Portugal. There are some areas of exposed sands and mud, reedbeds and saltmarsh. The surroundings consist mainly of open fields with small woods of oaks, pines, willows and alders.

The estuarine environment holds several interesting breeding species, like Marsh Harrier, Iberian Yellow Wagtail, Savi's Warbler, the introduced Common Waxbill and the rare *lusitanica* race of Reed Bunting. However, it is outside the breeding season that the estuary has greater potential for birding. The list of non-breeding visitors is long and comprises an array of migrant and wintering waterbirds, namely Little Egret, various ducks, waders and gulls – including some rarities at times. Seabird movement may be monitored from the mouth of the river and can prove rewarding. In the nearby fields and woodlands, it is possible to find several interesting landbirds, such as European Nightjar, Hoopoe, Woodlark, Fan-tailed, Melodious, Dartford and Sardinian Warblers, Iberian Chiffchaff, Firecrest, Crested Tit, Short-toed Treecreeper, Spotless Starling and Cirl Bunting.

The town of Caminha lies directly on the southern shore and can be used as a base for exploration. To get there from Porto, take the A28 motorway through Viana do Castelo to Caminha (100km drive).

From Caminha, the main road (N13) runs alongside the river and is convenient for exploring the estuary. To the south-west of the town lies Mata do Camarido – a small pinewood that ends right at the river mouth and gives access to the beach, from where the islet of Ínsua with its fortress can be seen. The pinewood itself is good for forest birds, namely tits and Firecrest; it is worth walking here. Leave Caminha on the N13 towards Viana and turn right where signposts read 'Foz do Rio Minho'.

To the east of Caminha the road leads to the bridge over the river Coura, a tributary which holds large expanses of reedbeds. To visit this place, one can either explore the left bank (by taking the N301 towards Paredes de Coura and stopping after 1 or 2km) or explore the right bank (from Caminha, take the road to Valença and when reaching Seixas, follow the signs to Vilar de Mouros; after 2km, veer right where signposts read 'praia fluvial'). Apart from marsh birds, this area is good for other small landbirds, such as tits, warblers and finches.

It is worth crossing the river Minho to Camposancos in order to explore the Spanish (Galician) bank. To get there, it is possible to take the ferryboat at Caminha (the price for a car is €3.00 as of 2013), but there is also a bridge at Vila Nova de Cerveira, about 12km to the north-east.

Serra do Gerês

The natural importance of this region was recognized in 1971 when it was declared as a national park, the only one that exists in Portugal (Parque Nacional da Peneda-Gerês). It extends for about 72,000ha and is particularly significant for its large and medium-sized mammals, which include Iberian Wolf and Roe Deer; more recently, the Spanish Ibex has been reintroduced. There is also a population of semi-wild horses, locally known as 'garranos'. The bird life is dominated by passerines and many northern Eurasian species reach their southern limits here.

Location *Geographical coordinates (Gerês village): 41°43'45.50"N, 8°09'45.00"W*

This mountainous area lies in the north-western corner of the country to the north-east of the town of Braga, which lies 50km north-east of Porto. Full details of the area can be found on the following maps: 1:100,000: 1, 2, 5 and 6; 1:50,000: 1-B, 1-D, 2-C, 5-B and 6-A; 1:25,000:4, 5, 9, 17, 18, 19, 30, 31 and 32.

Description This is a medium to high altitude region (500 to over 1500m), which includes not only serra do Gerês proper but extends over several other serras, specifically Peneda, Soajo and Amarela. Despite the proximity of the Atlantic, the temperature range is extreme because of altitude, resulting in hot summers and cold, often snowy, winters. The influence of a moist air stream off the Atlantic is responsible for the highest annual rainfall in Portugal.

There are some villages scattered throughout the area, although the most remote parts are uninhabited. On the higher hills there is little vegetation and the soil is often bare or simply covered with low scrub, while at lower altitudes there are some woods, consisting mainly of Pedunculate Oak and Maritime Pine, with a few Pyrenean Oaks, Birch, Scots Pine and Ash trees in some places. Around some villages, mainly in the eastern and northern sectors, the landscape forms a 'mosaic', typically with rye fields mixed in with some other northern crops. Disturbance has increased over the last thirty years, mainly due to tourism (a small fee is now payable in summer to drive through certain areas.)

Birds The common and interesting breeding birds include Black Redstart (which occurs in most villages), Firecrest, Crested Tit, Short-toed Treecreeper, Serin and Rock Bunting.

Other species, not so common but still to be found in suitable habitat include Quail, Scops Owl, European Nightjar, Hoopoe, Woodlark, Crag Martin, Iberian Yellow Wagtail, Fan-tailed, Melodious, Western Bonelli's and Dartford Warblers, Iberian Chiffchaff, Golden Oriole, Southern Grey Shrike, Chough, and Cirl and Ortolan Buntings. The open tops hold Rock Thrush, along with Tawny Pipit and Wheatear, while the mountain streams hold Dipper.

This area forms the Portuguese stronghold of several 'northern' Eurasian species, like Tree Pipit, Whinchat, Song Thrush, Garden Warbler, Red-backed Shrike, Bullfinch, Crossbill (irregular) and Yellowhammer.

The most common birds of prey include Buzzard, Montagu's Harrier (a proportion of which are all-dark morphs) and Sparrowhawk. Less common raptors include Black Kite, Honey-buzzard, Hen Harrier, Goshawk, Short-toed and Golden Eagles, and Peregrine Falcon.

In the colder months summer visitors are absent and many resident bird species undertake altitudinal movements, thus avoiding the

Rock Bunting

harsh conditions. At that time of year, the area has much less to offer. Nevertheless, Water Pipit and Alpine Accentor have been recorded in winter.

Accommodation

Finding accommodation should not be a problem, except perhaps in midsummer. At Gerês village there are several hotels and a campsite, while near the barragem de Vilarinho das Furnas there is a youth hostel. For those wishing to explore the eastern part of the park, it should be noted that there is a hotel at Montalegre, which lies further east, while in the northernmost part it is possible to find hotels at Castro Laboreiro and Melgaço.

There are many restaurants in the region.

Strategy

The best season for visiting this region is from April to August, when most breeding species are present. However, summer weekends especially in August should be avoided, because of the number of visiting tourists.

a) Central sector (Gerês village to Portela do Homem)

This is the most popular area within the national park. It is crossed by a winding road that goes from Caldas do Gerês (also known as Vila do Gerês) northwards to Portela do Homem, through Mata de Albergaria, which is a pristine oak woodland. Here it is possible to find Goshawk and Iberian Chiffchaff, along with several more common passerines. To get there, leave the N103 near the Pousada de São Bento and proceed through Gerês village on the N308-1 northwards to Portela do Leonte, where the best part of the woodland begins. A small fee (€1.50 as of 2013) is payable to drive through this area from June to September during daytime hours and parking is not allowed along the way, but pedestrian access is free of charge. Thus the best approach is to park at Portela do Leonte (southern end) or at Portela do Homem (northern end) and traverse the area on foot.

It should be noted that this place tends to be overcrowded in summer, so an early morning visit is essential in order to avoid the inevitable human disturbance.

b) Eastern sector (Mourela plateau)

The remote Mourela plateau lies within the eastern range of the serra do Gerês. It is much quieter than the central sector and definitely a better option for birding.

Take the N103 signposted to Chaves, which leads to Pisões (after 75km). Turn left on a small road just after Pisões, towards Covelães.

Start by exploring the area just before Covelães, including the road that leads westwards to Outeiro and Paradela. Interesting species to be found here include Honey-buzzard, Goshawk, Scops Owl, European Nightjar, Western Bonelli's Warblers, Golden Oriole and Rock Bunting.

After Covelães, continue northwards, stopping after 2km where the woodland gives way to more open country. always keeping an eye open for birds of prey. A few kilometres further north, the road forks. By taking the left fork, one reaches the small village of Pitões das Júnias, which lies at about 1100m. Park here and ask the way to the 'convento' (the old medieval convent), which can be reached in a 20 minute walk.

From Pitões das Júnias, there is an unsurfaced road that leads north-west towards Fonte Fria. Drive for a few kilometres, stop where the road ends and proceed on foot in the same direction; this walk brings you through some woods to the open tops. Montagu's Harrier, Western Bonelli's and Dartford Warbler and Rock Thrush are all possible here.

Returning to Pitões das Júnias, drive back to the fork and take the left

Serra do Gerês

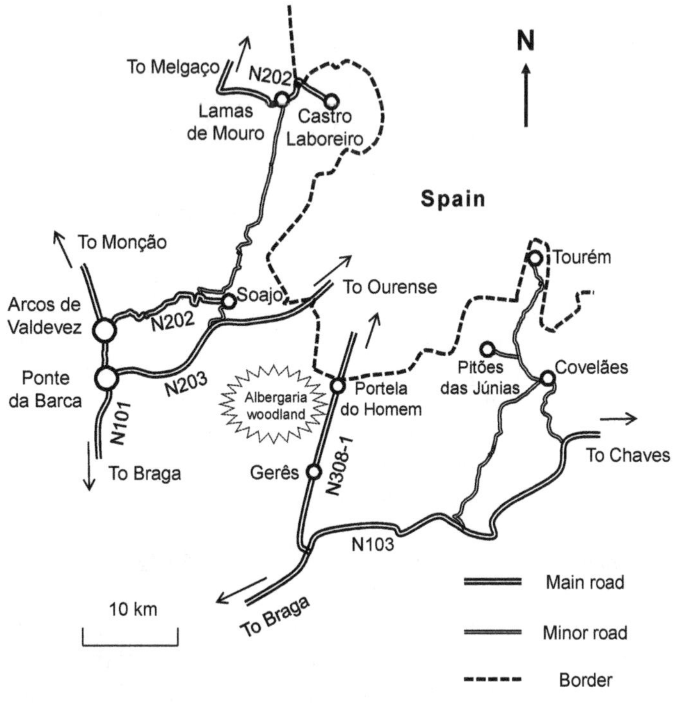

road to Tourém, which traverses the plateau (locally called planalto da Mourela). This is a good area for Red-backed Shrikes, which are common on the aerial wires here from mid-May onwards, and also for birds of prey. It is worth carrying on towards the Spanish border, beyond Tourém.

c) North-western part (Serra da Peneda)

From Braga take the A3 to Ponte de Lima, then the IC28 to Arcos de Valdevez and after that use the winding N202 road to Soajo (where the typical corn storage 'Espigueiros' should not be missed). From Soajo, a minor road leads northwards to Senhora da Peneda and then on to Lamas de Mouro. Alternatively, take the A3 to Valença and then follow the N101 and the N202 eastwards through Melgaço.

The area around Lamas de Mouro is a convenient base from which to explore. Here there is an information centre, where it is possible to get a map of the park, as well as detailed information about the places to visit.

There are several marked trails nearby which are good options for birding. One of these trails starts and ends at the information centre and goes through woodland, open hillsides and riverine areas. Along this 5km long trail it is possible to see many different passerines.

Further to the east lies the village of Castro Laboreiro. From here it is possible to access the higher altitude areas (above 1000m), which can be reached from the village of Portos, or take another marked trail, which goes through good patches of oak woodland. The most interesting species around here include Montagu's Harrier, Tawny Pipit, Rock Thrush and Ortolan Bunting, as well as several other common species which are also to be found at Lamas de Mouro.

When visiting this area, keep an eye out for dogs (the Castro Laboreiro, a race of Portuguese watchdog) which are sometimes allowed to run free near cattle herds and which can be dangerous if one approaches the cattle too closely. If in doubt steer clear of herds of cattle altogether.

Serra de Montesinho

Many central and northern European species reach their southern limit in serra de Montesinho – a chain of hills that has been declared a natural park – yet there are also many Mediterranean species present; the result is a remarkable mixed bird community. The park is in a mountainous region, which is hot in summer and very cold in winter. Human activities are mainly agricultural and there are no large centres of population in the park, but small villages exist here and there; for visiting birdwatchers, passerines are of most interest.

Apart from the birds, this area is well known for its populations of medium and large-sized mammals, namely Iberian Wolf, Red Deer, Roe Deer and Wild Boar.

Location

Geographical coordinates: 41°56'12.80"N, 6°45'18.00"W

Serra de Montesinho is situated in the north-eastern corner of Portugal, close to Bragança. The natural park's northern, western and eastern boundaries are formed by the adjacent Spanish border. The southern boundary is formed by the N103 (west of Bragança) and by the

N218 and N218-1, which run from Bragança eastwards to Quintanilha. Full details of the area can be found on the following maps: 1:100,000: 3; 1:50,000: 3-C and 3-D; 1:25,000: 10, 11, 12, 23, 24, 25, 26 and 39.

Description

The landscape is varied, complex and hilly, with most of the park at an altitude of over 700m. There are many different intergraded habitats, which occur throughout the region. The more open locations consist of cultivated and uncultivated plots, and occur especially in the eastern half of the park. There are many grassy fields, mainly used for grazing and pasture, which are divided by small stone walls or bushy hedges. These fields, locally called 'lameiros', are often surrounded by ash trees, Pyrenean Oak, poplars and birches.

Woodland is still an important habitat here, although much has been cleared. The park holds important groves of Pyrenean Oak, which originally covered almost the whole region, but are now restricted to those places which have not yet been transformed by more recent human activities. They are most frequent in the western and central half of the park. Many areas have been planted with a variety of pine species (Maritime, Scots and Corsican) and Douglas Fir. There are also many plantations of Sweet Chestnut, a species with a relatively high ecological rating in terms of associated biodiversity. Elms occur mainly in inhabited areas and along watercourses and roads. Unfortunately, practically all the Elms have died out in recent years, the result of a fungal disease which affected Iberia, along with many other European regions.

Birds

Passerines are undoubtedly the most interesting bird group to look for in this area. Common and widespread breeding species, which can be seen almost anywhere in the park include Tawny Pipit, Subalpine,

Subalpine Warbler

Western Bonelli's, Melodious and Dartford Warblers, Iberian Chiffchaff, Short-toed Treecreeper, Golden Oriole, Spotless Starling, Serin, and Cirl and Rock Buntings. Wrynecks are fairly common all over the park, best located in April and May, when their song can be frequently heard, usually around old Sweet Chestnut woods.

Less common species, which are rather more habitat dependent are: Short-toed Lark, Black-eared Wheatear, and Fan-tailed Warbler in open habitats; Red-rumped Swallow, Dipper and Blue Rock Thrush in deep river valleys; Cetti's Warbler in riverine woodland; Redstart and Rock Sparrow in Sweet Chestnut woods; Rock Thrush in the higher mountainous areas. The area holds three species of shrikes: Woodchat and Southern Grey Shrike are not uncommon, though much less numerous than further south, whereas Red-backed is reasonably common in higher places.

Water Pipit has bred here on the high ranges and can be seen in winter in small numbers. Crossbills occur in some years and may breed at times, mainly after irruptive years in northern or eastern Europe; they are usually seen in Scots Pine forests.

Birds of prey are not numerous. The only two common species are Buzzard and Montagu's Harrier. Rarer birds, which are more difficult to locate, include Honey-buzzard, Red and Black Kites, Egyptian Vulture, Hen Harrier, Golden and Short-toed Eagles, Goshawk and Sparrowhawk. Several pairs of White Storks breed here, as well as a pair of Black Storks.

Among the night birds, Scops is by far the most common owl and is easy to hear in the breeding season. Eagle Owl is very rare and finding one requires luck and patience; it begins its breeding cycle and thus its lugubrious song in early January. There are also European and Red-necked Nightjars, although the latter reaches its northernmost limit here and is very scarce. Other landbirds which can be found in the area include Quail and Bee-eater.

Winter can be very cold and sometimes snow covers the land. Birds are not numerous at that time of the year.

Accommodation

Bragança is the most convenient town in which to be based and there are several pensions and hotels to choose from. There are a few mountain huts scattered within the park that may be available through the park headquarters at Bragança. When using these, it is necessary to supply your own food, water and bedding.

There are a few restaurants in Bragança and in the villages of Vinhais and Gimonde.

Strategy

Late April to July is the best time to visit this place. Access is time consuming and two or three days may be necessary to explore the park thoroughly. A detailed map can be very useful.

There are many villages scattered throughout the area and these are served by a dense network of minor roads, which make it possible to explore in many directions. The following itineraries are some suggestions, but there are many other possibilities.

a) Eastern sector

The easternmost sector of the park can be explored by leaving

Bragança eastwards on the N218. After 4km, stop at Gimonde. Look for Crag Martin and Red-rumped Swallow, which breed under the bridge.

Proceed along the N308, until São Julião or Deilão are reached. This is an open, cultivated area, best explored on foot. Here it is possible to find some interesting species including Great Spotted Cuckoo, Short-toed Lark and Ortolan Bunting. Slightly further north, there are plantations of Scots Pine, where Crossbills are seen occasionally. Further south, near Quintanilha, it is worth exploring the valley of the river Maçãs, where there have been regular sightings of Egyptian Vulture and Goshawk; there are also many passerines around here.

b) Central sector

The central sector includes some montane locations, such as serra de Montesinho itself, which is excellent for high altitude species. From Bragança, drive north on the N103-7. Pass França and just before reaching the border at Portelo turn left to Montesinho. The road runs upwards and after about 4km one reaches a cross-roads.

Continue straight through, proceeding downwards until Montesinho village is reached; it lies in the bottom of a valley, at about 900m. The surrounding area, although highly agricultural, is very beautiful and is peppered with small woodlands of oak and birch. This area holds a variety of birds, including Golden Eagle, Wryneck and Red-backed Shrike.

If you turn right at the cross-roads mentioned above, you can proceed on an unsurfaced but negotiable road leading upwards to an almost treeless plateau at about 1300m above sea level, where there is a reservoir (barragem de Serra Serrada). From this point onwards there are good chances of seeing high altitude birds, such as Tawny Pipit,

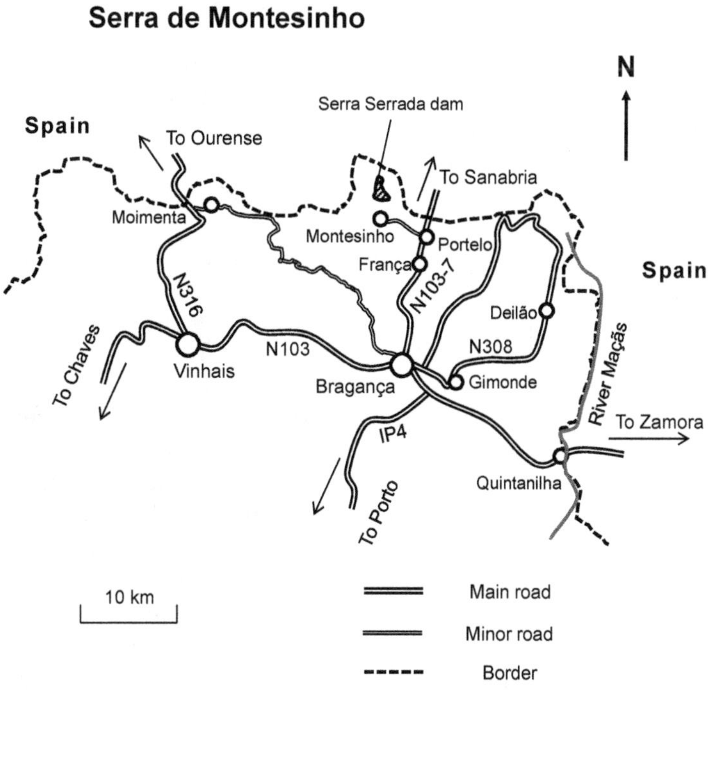

Rock Thrush, Red-backed Shrike, and Ortolan Bunting. A few more kilometres onwards, the road reaches the highest part of the park (over 1450m), which consists of high altitude pastureland and scrub, with a few Birch trees. The variety of birds is smaller here, but there are some specialities, including Hen and Montagu's Harriers, Rock Thrush, Southern Grey and Red-backed Shrikes, and possibly the only breeding Water Pipits in the country.

c) Western sector

In the western part of this region lies serra da Coroa, which is also integrated in the natural park. To get there, leave Bragança westwards on the N103 to Vinhais. From this town, a few roads lead northwards into the park, all of which are worth exploring. The best area however lies around Moimenta (25km to the north), near the international road to Ourense (Spain). It consists of a plateau where agricultural fields alternate with woods, pasture and scrub. Park the car near the village and explore the surroundings on foot. There is an enormous variety of breeding birds here, which include Honey-buzzard, European Nightjar, Red-backed Shrike and Ortolan Bunting, among many other common species.

International Douro

In the north-eastern corner of Portugal, the river Douro runs through a deep rocky valley. Apart from several dams that have been built there, the entire area is relatively unspoiled and a natural park was created in 1998, in order to protect it. This is a place particularly interesting for its birds of prey and for several species favouring rocky habitats.

Location

Geographical coordinates (Miranda do Douro): 41°30′00.00″N, 6°16′19.80″W

The river Douro, born in Spain, runs across the northern part of Portugal to Porto. The eastern sector, which forms the border between the two countries, is known as International Douro. Full details of the area can be found on the following maps: 1:100,000: 8, 11, 12 and 15; 1:50,000:8-C, 11-B, 11-D, 12-A and 12-C; 1:25,000: 67, 81, 94, 95, 107, 108, 120, 121, 131, 132, 142, 142A, 152 and 162.

Description

The area extends along the valley of the Douro from Miranda do Douro to Barca d'Alva. Its core is the valley itself, which is wide, cliff-lined and scrub-covered in some places, lending to the whole landscape a dramatic natural beauty. There are five dams along the river, mainly used for energy production, which have weakened the water current throughout much of its course. There are few buildings on the shore line, and human disturbance in the valley is relatively low.

Away from the valley, the vegetation becomes heterogeneous with open areas, consisting mainly of small to medium-sized cultivated and uncultivated fields, alternating with small woods. The dominant trees are Maritime Pines and Holm Oaks, usually surrounded by *Genista* scrub, but there are also Pyrenean and Portuguese Oaks, elm and ash.

Typical features of this region are the 'lameiros', small rectangular grassy fields (about 1 or 2ha in size), separated by small stone walls and mainly used for grazing. They are often surrounded by lines of ash trees.

Pyrenean Oaks and Sweet Chestnut groves are more common around Mogadouro and north of Miranda do Douro, while further south, around Freixo de Espada à Cinta and Barca d'Alva, the wooded areas consist largely of Almond trees and Olive groves.

Birds

This is one of the best sites for soaring birds in the country. Large raptors stand out, as this area holds the largest breeding populations of Griffon and Egyptian Vultures, as well as important numbers of Golden and Bonelli's Eagles. Other breeding species include Montagu's Harrier, Red and Black Kites, as well as Short-toed and Booted Eagles. There are also several pairs of White and Black Storks,

Owls include Scops Owl (numerous as a summer visitor) and Eagle Owl (still relatively common in suitable habitat). Additionally this area holds both European and Red-necked Nightjars.

Egyptian and Griffon Vultures

There is an abundance of breeding birds during the summer. Nightingales, Subalpine Warblers and Golden Orioles seem to be everywhere; other common summer visitors or residents include Alpine Swift, Bee-eater, Thekla Lark, Woodlark, Crag Martin, Red-rumped Swallow, Sardinian, Melodious and Western Bonelli's Warblers, Woodchat Shrike, Chough, Rock Sparrow, Hawfinch, and Cirl and Rock Buntings.

This area also holds some specialised or otherwise scarce species, including Great Spotted Cuckoo, Wryneck, Blue Rock Thrush and Western Orphean Warbler.

Accommodation

Miranda do Douro is a convenient place to stay. It is a rather small town with a few hotels and several pensions, most of them located near the centre. Some of these have river views so if you choose a room facing the valley it may be possible to birdwatch from the window. There is also a camping site near the town. 'Posta à Mirandesa', a particularly heavy, thick beef steak, is very typical dish and should definitely be sampled as there are several good restaurants here. Further south, accommodation can be found at Mogadouro and at Freixo de Espada à Cinta.

Strategy

The best time to visit this area is between April and July. Remember it can be almost unpleasantly hot in midsummer, especially between noon and 16.00.

This whole region is very remote, but the effort to reach it is definitely worthwhile, as the birdlife is so diverse and rich. The trip from Lisbon to Miranda do Douro can be made by car (the whole journey will take about 6 hours) or by bus (daily buses are run by Santos Viagens e Turismo and Rede Expressos – see the 'Useful Contacts and Websites' section at the end of the book).

a) Miranda do Douro

This picturesque and historical town lies close to where the river Douro enters the country. Since most birds can be seen within walking distance, the most sensible strategy is to use it as a base.

Start by exploring Miranda do Douro itself, with its beautiful monuments and splendid views over the river valley. Black Redstart, Spotless Starling and Rock Sparrow can be found in the cathedral tower or in the old castle; Blue Rock Thrush occurs on the cliffs and Golden Oriole wherever there are trees, even close to inhabited areas. Vultures and eagles fly over frequently.

To explore northwards, take the road to Ifanes and follow it for 3km, taking a minor road to the right. This leads to Vale de Águia. On arriving at this village, take one of the various tracks to the right that approach the Douro and explore on foot. This area consists mainly of large bushes and Holm Oaks, and it is possible to find Booted Eagle, Subalpine Warbler and Hawfinch here, while in the river valley it is possible to see large raptors, along with Alpine Swift and Chough. A good place from where to view the valley is the observation post called São João das Arribas, which lies close to Aldeia Nova, a small village 2km north of Vale de Águia.

Returning to to the road to Ifanes, proceed north and on reaching that village, veer eastwards to Paradela (the easternmost village in Portugal). On arriving at Paradela, take the furthest right fork which leads to the 'corner' of Portugal, where the river Douro enters the country. Here there are many uncultivated fields, as well as some Pyrenean Oak and Maritime Pine woods. Breeding species include, Wryneck, Thekla and Short-toed Larks, Woodlark, Subalpine and Western Bonelli's Warblers, and Hawfinch.

To the south of Miranda, the N221 is worth walking or driving for a few kilometres. A parallel track as far as Cércio is easier for walking. Here there are pinewoods and the ubiquitous 'lameiros', together with some open fields, always close to the river valley. Along this way it is not difficult to find Black Stork and large birds of prey, Bee-eater, Scops Owl, Golden Oriole, Western Orphean and Western Bonelli's Warblers, and Chough. Great Spotted Cuckoo has been seen in successive years just south of Sendim, which lies a few kilometres further south.

Evening and dusk can be very rewarding around Miranda do Douro, if the best sites are carefully and patiently explored. Red-necked Nightjars have been seen sitting on the avenues just outside the wall that surrounds the old town, while European Nightjars have been heard near the slope of the river. It is difficult not to hear Scops Owls from April to June in the trees just outside the town wall and sometimes even in the public gardens. To locate Eagle Owl, the best strategy is to listen for its calls, preferably after sunset. Try to locate two or three different points, not far from the rocky valley. Remember that these birds are scarce and patience is needed to hear and, especially, to see them. Also remember that they are more vocal in late December and January when their breeding season starts.

b) Mogadouro

This town lies in the middle of a large plateau and access from it to the Douro valley may take some time as it lies well to the west of the river.

The main road here is the IC5 (which was inaugurated in 2012) but for birding purposes it is probably a better option to drive along the 'old' N221, which has less traffic and has more places to stop in safety.

Starting from Mogadouro in the morning, take the N221 southwards for 26km, always keeping an eye open for soaring raptors. Large birds of prey are often seen flying over the road and occasionally even over Mogadouro town itself. After about 16km, there is a minor road to the left that leads to

Bruçó (signposted). This village is reached after 4km. Drive through the village and park the car. From here, it is possible to reach the river valley in a 20 min walk. This area is excellent for soaring birds, as well as for smaller, cliff-nesting birds such as Alpine Swift, Blue Rock Thrush and Chough. The *Genista* bushes on the way hold Dartford and Subalpine Warblers, while the Pyrenean Oaks along the road back to the N221 are worth exploring for Western Bonelli's Warbler and Golden Oriole.

Return to the N221 and drive for another 10km until the village of Lagoaça (signposted) is reached. It lies slightly off the main road to the east, close to the banks of the Douro. Go through the village and on reaching the graveyard ('cemitério' in Portuguese), take the right fork and proceed for about 500m until the track ends at a remarkable and breathtaking viewing point. Park the car and look for flying birds of prey, preferably around noon. Black Stork can also be found here. In the surrounding bushes and trees, Subalpine and Dartford Warblers, Golden Oriole, and Rock Bunting can be found, while Bee-eaters and Alpine Swifts are usually visible or audible overhead.

If staying at Mogadouro town itself, it is worth taking a walk around at dusk and listening for the characteristic calls of Scops Owls.

c) Freixo de Espada à Cinta

This small town lies directly on the slopes of the Douro valley, on the N220. There are two roads from the town worth exploring: for the first route, leave Freixo de Espada à Cinta southwards on the N220 and turn right after about 1km and then left at the next crossroads until you reach the top of a huge rock called Penedo Durão (point B on the map). Apart from affording excellent views, this is another good place for large soaring birds, Alpine Swift, and Chough.

Another possibility is to leave Freixo de Espada à Cinta southwards and keep on the N220. Pass along a very winding road that gets ever closer to the valley and crosses the river at Barca d'Alva. An impressive line of cliffs can be seen on your right hand side and it is worth stopping here, looking for soaring vultures and eagles or even Peregrine Falcons. Azure-winged Magpies are not difficult to find here; close to the Douro, there are also Bee-eaters, and Golden Orioles, while the craggy tops hold Black Wheatears.

A few kilometres east of Barca d'Alva (on the northern side of the Douro) a small road leads to a rocky valley called ribeira do Mosteiro (point A on the map). This is a very beautiful area, where it is possible to find several cliff-nesting birds. This is another good site to look for the elusive the Black Wheatear, which often perches on top of the rocks. For those who enjoy walking, the 'calçada de Alpajares' (an old Roman road) in the vicinity is also an interesting possibility.

Cávado Estuary

Geographical coordinates: 41°31′56.30″N, W 8°46′50.50″W

A small wetland area lying close to Porto, the Cávado estuary can be visited in a short day's trip from this large town. It consists mainly of exposed mud and is surrounded by some agricultural fields and a few woods. A small hide has been built here. The estuary is separated from the sea by sand dunes, which form a line along the beach.

Like most estuaries in Portugal, Cávado is noted for its shorebirds – Kentish Plover is the only breeding species, but a much larger variety occurs on passage. Other interesting waterbirds which can be regularly found outside the breeding season include Black-necked Grebe, Little Egret, Marsh Harrier and Osprey. A few rare birds have been recorded here, namely divers, waterfowl and gulls.

In the surrounding areas it is possible to find several landbirds typical of southern Europe, such as Fan-tailed and Melodious Warblers, Firecrest, Spotless Starling and Cirl Bunting.

To get here from Porto, take the A28 northwards and leave it after 40km, following the signs to Esposende – this town lies next to the estuary.

The right shore can be easily explored by following Esposende's main avenue, which runs along the estuary, all the way to the beach. The left bank is quieter and is also worth a visit, especially in the afternoon, when sunlight is more favourable; to get there, leave Esposende on the N13 following the signs to Porto and cross the bridge to Fão, veering right on the small road to Ofir. From Fão, a boardwalk has been built along the southern shore and leads to a hide, providing good views over the estuary.

Porto and Surroundings

Geographical coordinates: 41°08'30.00"N, 8°39'41.00"W

The second most important city in the country, Porto is the capital of northern Portugal. It lies near the coastline, about 300km north of Lisbon, on the north shore of the river Douro. Just south of Porto lies the town of Vila Nova de Gaia which has grown markedly in recent years. Both towns are separated by the river Douro, one of the largest Portuguese rivers, which here forms a narrow estuary before entering the sea.

The Porto area provides the visiting birdwatcher with several southern species and although many Mediterranean species are largely absent it is nevertheless possible to find an interesting array of resident and migrant birds – about 280 species have been recorded in Porto district alone.

Breeding species that are typical of Iberia and that can easily be found in or around Porto include Fan-tailed, Melodious and Sardinian Warblers, Firecrest, Short-toed Treecreeper, Spotless Starling and Serin.

In terms of breeding waterbirds, the place is not especially rich, however during migration times and in winter the estuary is full of them, notably waders and gulls, along with Little Egret, Spoonbill and Bluethroat. A few Nearctic rarities have been found in the area, including Lesser Yellowlegs and American Herring Gull.

The best site to look for waterbirds is the Douro estuary itself, especially on the southern shore, where a local nature reserve has been declared – the reserve is run by the Vila Nova de Gaia municipality. A wooden hide has been built and provides a good view over the estuary. The easiest way to reach this area is to cross the river southwards using the Arrábida bridge and then leave the motorway following the signs to Afurada, which lies on the southern shore, close to the river mouth. For those interested in gulls, it might also be worth exploring the northern shore west from Porto towards Matosinhos or even further north at Leixões harbour. To get there, look for the large Avenida da Boavista (the longest in Porto) and follow it westwards until you reach the sea at Castelo do Queijo. From here, one can either proceed southwards to the Douro river mouth or northwards to the harbour.

Landbirds can be found at two main sites, one in either town: in Porto it is worth visiting the Parque da Cidade ('city park'), which lies in the western part (near Matosinhos) and is the largest green area in the town – this place can easily be reached by public transport, namely Metro do Porto – look for the line that runs to Senhor de Matosinhos; in Vila Nova de Gaia look for the Parque Biológico de Gaia, which lies just east of Vila Nova de Gaia, near Avintes. Yet another interesting place is the park of the modern art gallery of Serralves – it lies in the western end of Porto, just south of Avenida da Boavista and can easily be reached by bus.

CENTRE

Ria de Aveiro

The term 'ria', used now for many years, describes the geomorphological development of a coastal delta-like formation, unlike the 'rias' of geological origin found further north in Europe. Some authors describe such modified deltas as haff-deltas, something particularly accurate where Aveiro's labyrinthine bay is concerned.

Formerly one of the most important estuarine wetlands in Portugal, the Ria de Aveiro has lost much of its ecological value due to a growing human population, heavy industrial pollution, and the abandonment of salt works with subsequent conversion into fish ponds. Illegal hunting and legal over-hunting have also affected the area. Nevertheless, it still holds the best selection of waterbirds in northern Portugal, including most of the species one would expect to find in coastal wetlands in south-western Europe. Moreover, it boasts a selection of interesting landbirds typical of southern Europe.

Location

Geographical coordinates: 40°38'36.00"N, 8°39'38.70"W

The city of Aveiro lies about 60km south of Porto and 240km north of Lisbon, on the west coast of Portugal. The ria itself lies to the north and west of the town. Full details of the area can be found on the following maps: 1:100,000: 13 and 16; 1:50,000: 13-C and 16-A; 1:25,000: 163, 173, 174, 184 and 185.

Description

Aveiro's large wetland area is about 45km long and 12km wide, covering about 11,000ha, of which 6,000ha are non-tidal and over 1,000ha are saltpans. The water depth varies due to dredging and is generally low, except in the main channels, locally known as 'cales'. The intertidal vegetation is varied and ecologically quite interesting and important. Sand and mudflats are also extensive, the latter being particularly rich in algae.

The salt extraction business is now much reduced and this has led to increasing saltpan desertion and to consequent decay of the banks and tanks. The remaining complexes have bare, grassy or bushy hummocks, tamarisk lines and some reedy ditches. As has happened elsewhere in Portugal, many of the old saltpans have been converted into fish ponds.

The lowland between the rivers Vouga and Antuã consists of an 'enclosures' landscape. This changes gradually as one proceeds through the area to the north-west from small meadows and pastures, surrounded by tall hedges, to more open and wet ground with ditches, reedbeds, rushes, tamarisk and rice fields.

Along the northern sandbar, there are Maritime Pine and acacia woods that were planted by the Serviços Florestais (Forestry Services) between 1888 and 1924 to protect the dunes system. Part of this woodland and dunes is designated as a nature reserve (Reserva Natural das Dunas de São Jacinto); it lies between São Jacinto and the Orbitur camping site. Some decades ago this was mainly a well preserved pinewood, but fires in recent years have led to the partial destruction of the former woodland and large parts are now dominated by acacia trees. The reserve was originally declared with the purpose of protecting a heronry which no longer exists. However, two ponds have been created within the reserve and these have proven effective as a wildlife refuge, particularly for waterfowl.

Birds

The Ria de Aveiro is good for birds at any time of the year. However, like most Portuguese wetlands, it is more exciting during migration periods, particularly in autumn, when a wide variety of migrants pass through, and during winter, when large number of waterbirds gather in the mudflats, the saltpans and the neighbouring water bodies.

Great Reed Warbler

Waders are usually plentiful, with most western European species being present in the area outside the breeding season. The most abundant are Dunlin, Ringed Plover and Black-tailed Godwit, but many other species turn up regularly, including Avocet, Little Stint and Greenshank. During the breeding season the dominant wader species are Black-winged Stilt and Kentish Plover.

Gulls are also abundant; Mediterranean Gulls which were rare in the past are nowadays a regular sight. Little Gull occurs on occasion and rarer species, like Ring-billed and Glaucous Gulls have been recorded either in the Ria or in nearby places.

Waterfowl tend to concentrate at the São Jacinto reserve, because of hunting pressure in the surrounding areas. Wigeon and Teal are usually the dominant species, but many other species put in an appearance. There have been a few records of Red-crested Pochard, Ferruginous Duck and even American Wigeon and Green-winged Teal.

Other large waterbirds that can be seen outside the breeding season include: Little Egret, Squacco Heron (scarce), Spoonbill, Glossy Ibis and Greater Flamingo.

The Ria and the surrounding areas also attract birds of prey. The most regular species are Sparrowhawk, Buzzard and Marsh Harrier, which are resident. Black Kite turns up in summer, while non-breeding visitors include Black-winged Kite, Osprey, Merlin and Peregrine Falcon. Short-eared Owl also occurs in winter.

The reedbeds and the riverine woodlands in the lowland areas attract several breeding birds, of which the following are noteworthy: Purple Heron, Cetti's, Savi's, Reed, Great Reed and Melodious Warblers.

Finally, the whole place holds many passerine species. Wintering birds include Crag Martin, Water Pipit and Bluethroat, while Black Redstart, Fan-tailed, Dartford and Sardinian Warblers, Firecrest, Crested Tit, Spotless Starling and the exotic Common Waxbill are resident; Hoopoe and Iberian Yellow Wagtail are summer visitors. All these species are quite common in suitable habitat. Other breeding birds include White Stork and Pallid Swift (the latter breeds in Aveiro and is easy to spot in spring and summer around the town centre).

Other interesting species which can be harder to find but are still regular in the area, include Little Bittern, Goshawk, Great Spotted Cuckoo, Short-toed and Crested Larks, Woodlark and Cirl Bunting.

Accommodation

Aveiro has a variety of accommodation ranging from luxurious hotels to rented rooms. It also has a number of good restaurants. Another possibility is to stay on the northern sandbar; at São Jacinto the choice is limited, but there is a camping site (where 4-person bungalows are available) and a few kilometres to the north, at Torreira, there are other options.

Strategy

To explore the area thoroughly, it is probably best to base oneself at Aveiro. This town can easily be reached from Lisbon or Porto by taking the A1 motorway as far as Sobreiro junction (exit No. 16) and then following the A25 which runs around Aveiro. From Aveiro one can either explore the saltpans (all within walking distance of the city centre) or drive around the Ria.

a) Aveiro saltpans

Access to some saltpans ('salinas' or 'marinhas' as they are locally known) is usually difficult, as many of them lie on islets and can only be reached by boat. Luckily, there are three main complexes with access from land:

Marinha da Troncalhada – this is one of the most accessible saltpans and also one of the best for shorebirds. An open air museum about salt ('Ecomuseu do Sal') exists here. Follow the main city channel ('Canal das Pirâmides') north-west on the left bank, passing below the A25, until the saltpans become visible on the left hand side. The sign 'Marinha da Troncalhada' indicates the access. Explore this area thoroughly, as this is usually the best place for small waders. Other birds here can include marsh terns (on passage) and Iberian Yellow Wagtail.

Marinhas de São Roque – this complex lies just north of the town, across the São Roque channel. Follow the right bank of the main city channel which runs through the city from Rossio, passing under the A25. Proceed to the fishing harbour and then turn right for the saltpans. Alternatively, cross the São Roque channel at its western end and drive along the parking lot until you reach a raised pedestrian crossing. Here you can cross the motorway directly into the saltpans. These are very good for small waders and other waterbirds, including egrets and grebes. Raptors are often present and the bushy hummocks hold several passerines, like Dartford Warbler and wintering Bluethroat.

Marinhas de São Tiago and Lago do Paraíso – this site is located at the edge of the urban perimeter of Aveiro and can easily be reached, even on foot. Leave the car at Rua da Pega (which runs between the University and the Ria) and explore by walking. These saltpans are one of the best wintering areas for Avocet, Little Egret and Crag Martin, while Marsh Harrier and Peregrine Falcon can often be seen hunting here. In the saltpans there are breeding Kentish Plovers. To the south-east, there are reedbeds, tamarisk and rushes with Cetti's, Great Reed and Melodious Warblers and Common Waxbill; Bluethroat and Dartford Warbler winter here. The mudflats can also be interesting at low tide.

b) The lowlands at Salreu

The lowlands between the rivers Vouga and Antuã valleys are extensive with many bird-rich sites, only the most interesting and accessible of which are outlined here. One of the best places is at Salreu. From Aveiro, follow the A25 as far as Cacia and then turn onto the N109 to Salreu. At Salreu turn left, crossing the railway line with care. Just after traversing it, stop at the Environmental Interpretation Centre. Here it is possible to get information about the nature trails and the best places to visit. This centre is part of the Bioria project, which is run by the local municipality (Estarreja). As part of this project, a network of paths totalling 40km has been created. More information about the existing paths can be found at the project website (see the 'Useful Contacts and Websites' section at the end of the book).

After leaving the centre, proceed westwards as far as the track allows. Park the car and walk in the same direction until you reach the river Antuã (if possible, take a different track on your way back to the car).

Marsh Harriers and Purple Herons are common in the larger reedbeds, which are also the best areas for Little Bittern and Savi's Warbler. Great Reed Warblers seem to prefer reedy ditches, while Great Spotted Cuckoo and Melodious Warbler can be found mainly on the larger hedges. In autumn some unusual migrants have been recorded (including Squacco Heron, Red Kite, Hobby and Red-rumped Swallow). Spoonbill is seen on passage and may overwinter.

Another excellent area is Vilarinho/Sarrazola, which lies further south, about 2km north-west of Cacia. Leave Aveiro towards north-east on the N109 and after 4km, on reaching Cacia, turn left towards Vilarinho and Sarrazola; the road passes over the railway. On reaching the next T-junction, turn left, on the next crossroads continue straight ahead, on the following one turn right and proceed to the Sarrazola concrete bridge, from where it is possible to park and walk following the marked paths. Common species in this area include Hoopoe, Cetti's Warbler, Common Waxbill and Cirl Bunting, while less common birds include Little Egret, Black-winged Stilt, Crested Lark, Water Pipit, Firecrest and Golden Oriole, depending on season.

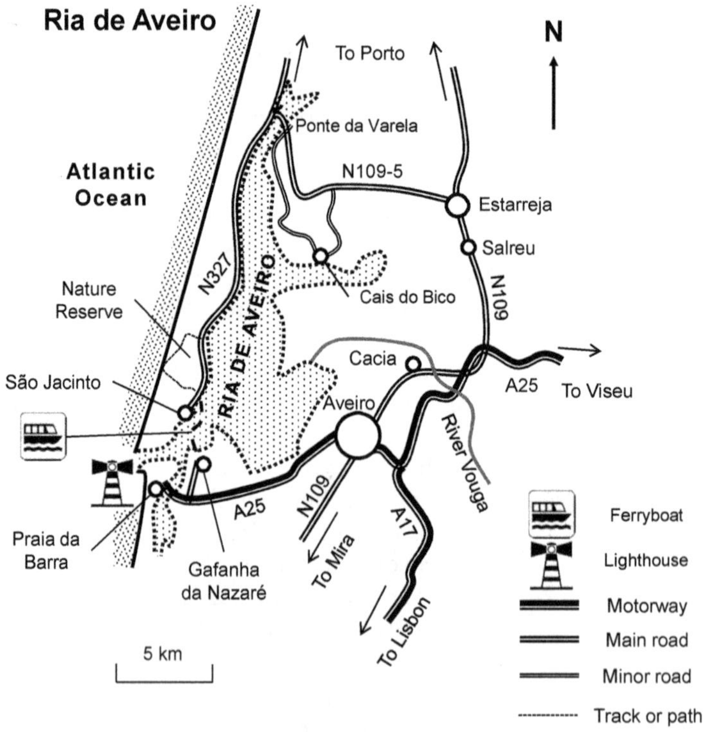

c) Northern sandbar and Murtosa area

Access to the northern sandbar can be made by car, either using the ferryboat to São Jacinto or driving all the way around the Ria through Estarreja and then westwards on the N109-5 until the ponte da Varela ('Varela bridge') is reached – this bridge crosses the channel to the sandbar. Probably the best approach is to do the complete circuit (i.e.

take the ferry and then drive back through Estarreja, or vice-versa). The following is a description of the circuit done in a clockwise direction.

From Aveiro take the A25 westwards and leave it at exit No. 1, signposted 'Gafanha da Nazaré'; follow the signs to the ferryboat. The crossing to São Jacinto takes about 15 minutes. As of 2013, the tickets cost €2 per passenger and €8 per car, including driver (please see the 'Useful Contacts and Websites' section at the end of the book for timetables). Crested Lark is often seen on the roads leading to the ferryboat. During the boat crossing, keep an eye on the gulls or other waterbirds.

Upon arriving at São Jacinto, look for the nature reserve (Reserva Natural das Dunas de São Jacinto). The reserve headquarters lie directly beside the N327, just north of São Jacinto settlement, and can provide useful information to visitors (it is closed on Sundays). There is a marked trail which goes through the woodlands and which passes near two man-made ponds (with hides) that hold good numbers of wintering waterfowl. Goshawk and Sparrowhawk are also regular here, while Osprey may occur in winter.

From São Jacinto the N327 runs northwards. This road offers a unique view over São Jacinto channel and the adjoining islets which lie to the east (particularly between Muranzel and São Jacinto). There are good mudflats at low tide and some open water which is always worth checking.

After about 15km, one reaches the large Varela bridge, which links the sandbar to the Murtosa peninsula. The best way to explore this area is to drive along the shore. To do this, cross the bridge and then turn left, following the signs to 'Bestida' and then to 'Cais do Bico'. A narrow road runs along the Ria and provides good views over the mudflats; waders are plentiful during low tide and other waterbirds are often present, including Little Egret and sometimes Greater Flamingos. The fields on the left shouldn't be neglected – Short-eared Owl occurs in winter.

From Cais do Bico proceed north-eastwards to Murtosa and then eastwards to Estarreja.

Lower Mondego (Marshes and Estuary)

Geographical coordinates (Mondego estuary): 40°08'20.80"N, 8°50'25.70"W

The river Mondego runs from serra da Estrela, passes through the historic city of Coimbra and enters the sea at Figueira da Foz. Between these two cities, it flows across a wide flat valley now occupied mainly by extensive rice fields. The remaining natural wetlands are small but they hold an interesting bird community. The entire area lies conveniently close to Coimbra and to Figueira da Foz.

a) Mondego marshes

There are a few marshy wetlands dotted over the lowland, the main ones being Arzila, Taipal and Madriz. All these places have a dense growth of aquatic vegetation, especially reeds.

Interesting species to be found in the breeding season include Purple Heron, Little Bittern, White Stork, Black Kite, Marsh Harrier, Hobby, Reed, Great Reed and Savi's Warblers and Golden Oriole. Wintering species include various ducks, Hen Harrier, Lapwing and Crag Martin, while Bluethroat and Sedge Warbler occur on migration. Squacco Heron, Great White Egret and Glossy Ibis have been seen here.

Paul de Arzila is not far from Coimbra and access is easy. From the town, take the N341, which follows the railway line, along the left bank of the Mondego and proceed for 11km. At Arzila village, turn left where the signpost reads 'Condeixa', 'Cemitério' and 'Paul', and then to the right following the signs to 'Paul'. This leads to Paul de Arzila, which is a nature reserve. There is a visitor centre nearby. To explore the site, just follow the marked paths.

Paul da Madriz lies close to Alfarelos and can be reached by proceeding on the N341 from Montemor-o-Velho to Alfarelos, turning left where the signpost reads 'Soure' and continuing along the N342-1, past Casais do Redinho for another 600m, until a valley with willows and reeds is reached. The marsh lies on the right side of the road.

Lastly, Paul do Taipal lies just west of Montemor-o-Velho and is crossed by the N111, which connects Coimbra to Figueira da Foz. There is a hide close to the road.

b) Mondego estuary

The estuary at Figueira da Foz consists mainly of sandflats and intertidal areas, although there are also some saltpans in an island in the middle of the river (called Morraceira). Black-winged Stilt, Kentish Plover and Iberian Yellow Wagtail breed, while passage brings large flocks of waders and gulls. Greater Flamingos are a regular sight here.

The town can be reached using the A8 and A17 motorways from Lisbon or the A14 from Coimbra. From Figueira da Foz, cross the river on the main bridge and turn right where the signpost reads 'Estaleiros' and then again to the right, taking a narrow well surfaced road that leads into the saltpan complex.

Ribacoa

Ribacoa region is one of the least known yet most well preserved areas of mainland Portugal. This plateau, lying at 800 m height, extends along the eastern side of the river Coa and holds an exceptionally rich birdlife. For the visiting birder, one day or two here can provide an incredible birding experience, as most Mediterranean bird species are present and can easily be found.

Location

Geographical coordinates (Vilar Formoso): 40°36'39.00"N, 6°49'47.60"W

This region lies in eastern Portugal, between the Douro valley and the serra da Malcata, along the Spanish border, 350km due north-east of Lisbon. Full details of the area can be found on the following maps: 1:100,000: 15 and 18; 1:50,000: 15 – B, 15 – D, 18 – B and 18 – D; 1:25,000: 152, 162, 172, 183, 194, 205, 216, 226 and 227

Description

Because of its altitude, this area bears a resemblance to the more remote province of Trás-os-Montes in the north-east but it is more accessible and easier to reach.

The site can be divided in two parts: the northernmost part (from Vilar Formoso north to Almeida and Figueira de Castelo Rodrigo) is a rather dry plateau, with scattered Holm Oaks and small pine plantations alternating with more open areas. The reservoir of Santa Maria de Aguiar is the largest in the region.

Further south, towards Sabugal, the woodland gets denser, with Pyrenean Oak being the dominant species. The serra da Malcata, the southern limit of the area, has a very low human population and has become famous as one of the last places for Iberian Lynx in Portugal – this mammal is now very rare and it is not certain if it still occurs here.

About 15km west of Figueira de Castelo Rodrigo, on the river Coa valley, there is a small nature reserve (Faia Brava); this is a private protected area, which is owned and managed by Associação Transumância e Natureza (ATN), a local NGO.

Lying close to the border with Spain, this is a region with a long history of wars and battles. It is thus not surprising that there are many castles, fortresses and other historical remnants from the middle ages. In Sabugal county alone there are five different castles, all made of granite blocks. Further to the north, at Almeida, there is an impressive star-shaped fortress from the mid-17th century. Throughout the area there are many villages with typical houses made of stone. Some of these villages, such as Almeida, Sortelha, Castelo Mendo or Castelo Rodrigo, have been redesignated for tourism purposes and are now known as 'Aldeias Históricas' (historical villages). All these castles and villages are worth visiting, both for their history and for their birds,

Agriculture is still a major occupation, people working in the fields a frequent sight. Rye and wheat are the main crops cultivated around the area, but there are also a few Sweet Chestnut woods.

Birds

This is one of the richest areas, bird-wise, in the whole country, especially in terms of soaring birds and also passerines.

Among the former, vultures are often present and they breed in the northern part of the area – but it is also likely that many birds seen here are drifting individuals from nearby Spanish colonies. Other species include Black-winged, Red and Black Kites, Montagu's Harriers, Short-toed, Golden, Bonelli's, and Booted Eagles, Honey-buzzard and Hobby. White and Black Storks are also found regularly. Scops Owl can be heard singing in spring in several villages throughout the region.

As for the passerines, the diversity of forest birds should be highlighted. The oak woodlands are frequented by Nightingales, Western Bonelli's, Melodious and Subalpine Warblers, Short-toed Treecreepers, Woodchat Shrikes, Golden Orioles, Azure-winged Magpies and Rock Buntings. In the more open areas it is also possible to find a number of interesting species, notably larks. Calandra Lark, a rare species in northern Portugal, seems to be regular here, along with Woodlark, and Thekla, and Short-toed Larks. Spanish Sparrow also occurs in the area, although it is far less common than in the southernmost provinces.

The villages, where stone houses are in the majority, are ideal places to look for Black Redstart, Rock Sparrow, Spotless Starling and sometimes

Blue Rock Thrush, while Crag Martin and Red-rumped Swallow can easily be found throughout the region, often around small bridges.

Accommodation

Sabugal and Vilar Formoso have at least one hotel each and several hostels. There are more hotels in the nearby town of Guarda (40km to the west). More familiar 'rural tourism' can be found in Sortelha and at a few other places throughout the area.

Strategy

Spring is definitely the best season to visit this region – this is the time when most song birds can easily be found in the surrounding woodlands.

The easiest way to reach the area from the coast is on the A25 motorway, which runs east-west and intersects the main A1 (Lisbon-Porto) near Aveiro. The A25 leads to the border village of Vilar Formoso (where it crosses into Spain). From here it is possible to explore northwards or southwards.

a) South towards Sabugal and Serra da Malcata

From Vilar Formoso (where Black Kite is easy to see flying over the streets), take the N332 southwards – this road provides access to most interesting places and eventually leads to Sabugal, a small town that is conveniently located to explore the southern part of the area. There are several villages in the region and the road network is quite reasonable providing access to most places.

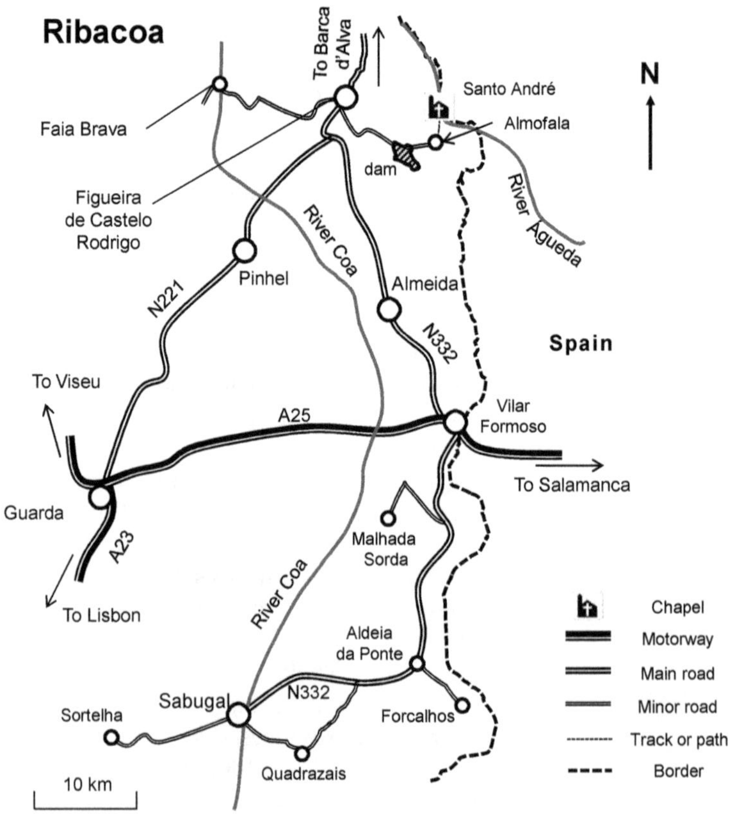

Just to the south of Vilar Formoso, one enters an open area, suitable for looking for open country birds. It is best to use secondary roads, as the main one often has heavy traffic. The narrow road to Malhada Sorda crosses some open fields which are favourable for larks. The scattered Holm Oak woods are good for Western Orphean Warbler but this species occurs in low densities, though, and can be difficult to find.

Further south, one reaches Aldeia da Ponte – around here and on the way to the nearby small village of Forcalhos, 3km to the south-east, lies one of the finest places for birds. The landscape is dominated by good patches of Pyrenean Oak, although pine and ash are also present. The surrounding oak woodlands are good places to look for Western Bonelli's Warbler and other passerines.

The rivers and streams in the region hold Dipper, for example between Sabugal and the village of Quadrazais.

West of Sabugal, the small historical village of Sortelha is also worth a visit, both for the breathtaking views and for its birds – Crag Martin, Red-rumped Swallow, Blue Rock Thrush and Rock Bunting, are all easy to spot from the castle, which is also a good place to scan for flying raptors around mid-morning.

b) North towards Figueira de Castelo Rodrigo

This area can be explored following the N332 which connects Vilar Formoso to Almeida (15km) and Figueira de Castelo Rodrigo (40km). Some of the most interesting points to visit are mentioned below.

The town of Figueira de Castelo Rodrigo has breeding White Storks and in the trees of the main square it is usually possible to hear the song of Scops Owl in spring.

South-east of this small town, between the villages of Nave Redonda and Almofala, the reservoir of Santa Maria de Aguiar is the most important wetland in the region and is worth checking – in winter there are usually some waterfowl, while in spring Black Kite can often be found patrolling the water's edge. In the surrounding areas it is possible to find other raptors and also many passerines, including Azure-winged Magpie. Further east, on reaching Almofala village, look for the small unsurfaced road that leads northwards to Santo André das Arribas – this is a small chapel with an outstanding view over the river Águeda. This deep valley is good for cliff-nesting birds, namely vultures and large eagles, so it is worth spending some time scanning the rocky gorges. In the scrubby areas it is possible to find Subalpine and Melodious Warblers along with several other species of passerines.

The Faia Brava nature reserve also has several species of cliff-nesting birds. Here there are several trails and there is also a hide which can be used to photograph vultures at close range (a fee is payable). More details about this place can be found on the ATN website (see the 'Useful Contacts and Websites' section at the end of the book).

Serra da Estrela

This mountain chain is the highest in mainland Portugal, peaking at 1993m (a 7m tower has been built on the top, in order to reach the

2000m mark). It is popular both in summer and in winter as a tourist resort. Apart from the magnificent scenery, this area is probably the best place in Portugal to watch specialised species of montane birds.

Location

Geographical coordinates (Torre): 40°19′18.60″N, 7°36′47.00″W

Serra da Estrela lies in inner central Portugal, some 300km from Lisbon. Full details of the area can be found on the following maps: 1:100,000: 17 and 20; 1:50,000: 17-D, 20-B; 1:25,000: 212, 213, 223, 224 and 235.

Description

Based primarily on altitude, the area can be divided into three main levels. The lower one (sometimes called *basal*) lies below 1300m and is occupied by Sweet Chestnut groves, as well as pine plantations, and some agricultural fields (where rye is cultivated). There are a few scattered villages.

Above 1300m lies the second level, called *subalpine*; here, there are large extensions of bushes, mainly *Genista*, and a few plantations of Scots Pine, Corsican Pine and Birch; several medium size dams have been built throughout the area, mainly for power generation, and there are a few natural lakes; rocky outcrops are commonplace. At this level there are no permanent human habitations, but there is an important tourist resort at Penhas da Saúde, on the eastern side. The glacier valley of the river Zêzere runs northwards and its most beautiful part lies at this level.

The topmost level, above 1600m, is largely bare rock only; the vegetation consists mainly of small Junipers and grass. The scenery is spectacular with far-reaching views. A tarmac road leads right to the top, where there is a restaurant and a small shopping centre. A visitor centre is located right next to it. Another visitor centre – Centro de Interpretação da Serra da Estrela (CISE) – is located at Seia, the main town on the north-western slope.

Birds

As can be expected in such a mountainous area, the number of species is higher at lower levels and decreases as one moves uphill. However, close to the top a few special birds can be found.

In the lower areas at around 1000m, birdlife is still quite varied, although already dominated by species that are adapted for a montane climate. Interesting birds to be found at this level include Woodlark, Crag Martin, Black Redstart, Serin and Rock Bunting. The towns and villages hold Swift and Pallid Swift, and Red-rumped Swallow can be found locally around buildings on the hillsides.

Between 1300 and 1600m one enters the submontane area. This is a good place for woodland birds, such as Western Bonelli's Warbler, Firecrest, Coal Tit and Short-toed Treecreeper, whereas the more open areas are frequented by Dunnock, Dartford Warbler, Whitethroat and Rock and Ortolan Buntings. The mountain streams hold Dipper and Grey Wagtail.

Above 1600m there are no trees and the birdlife is largely dominated by half a dozen species in the breeding season: Skylark, Tawny Pipit, Dunnock, Black Redstart, Wheatear and Linnet are commonplace. On the more rocky parts, Crag Martins fly overhead and Rock Thrush can be heard singing in spring. Now and then a small flock of Choughs turns up, but the species no longer breeds in the area. In winter there are very few

Rock Thrush

birds here, but Ring Ouzel, Water Pipit, Alpine Accentor and Snow Bunting are all regular.

Birds of prey can turn up at any place, although in small numbers. Among the more regular species are Black Kite, Short-toed and Booted Eagles, Montagu's Harrier, and Peregrine Falcon, but occasionally other species turn up, such as Honey-buzzard and Hobby.

Accommodation

Due to the existing tourism facilities at Manteigas, Sabugueiro, Covilhã, finding a place to stay is not a problem, except possibly at critical times such as Carnival, Easter and New Year and in general on winter weekends, when ski fans turn up in big numbers. Seia, on the western side, and Fundão, near serra da Gardunha, are other possibilities.

Strategy

The most interesting season to visit serra da Estrela is during the breeding season, which takes place here a little later than elsewhere in the country, because of altitude. May to July are probably the best months. However, interesting birds can also turn up at other times of the year, even in the cold winter months.

To get to this region from Lisbon, it is best to follow the A1 to Torres Novas, then travel onwards on the A23 to Covilhã (alternatively, one can follow the A1 to Coimbra, then the IP3 and IC12 to Nelas and finally the N232 to Seia).

The roads that run uphill are very winding and in many places it is not possible to stop. Fortunately there are a few lay-bys which allow safe parking and these make it possible to explore and search for the most interesting species.

The serra can be approached from the eastern side (via Covilhã) or from the western side (via Seia). The two routes are described here.

a) Eastern side

Covilhã town is conveniently located for exploring the area. Leave the town following the signs reading 'Serra da Estrela'. As one proceeds uphill, there are several signs along the road which indicate the altitude. After passing the 1000m mark, look for a hotel called 'Varanda dos Carqueijais'. This point gives breathtaking views to the east and is a regular spot for Rock Bunting. A few kilometres ahead the mountain resort of Penhas da Saúde will appear. Here it is possible to park and look for passerines. Birds around here usually include Black Redstart, Firecrest, Serin and sometimes Rock Thrush, while Ring Ouzel has been recorded in winter.

Proceed uphill until you reach a crossroads – this place is called Piornos (not signposted but easily recognizable because there is a large park full of snowploughs). A small road to the left leads to an abandoned cable car building. Park the car and walk around. This place is excellent for passerines in spring, including Tawny Pipit and Ortolan Bunting. In recent years Spectacled Warbler has been recorded here.

From Piornos it is possible to take the road uphill towards Torre (see below, under 'western side'). Another option is to veer right on the road to Manteigas, which follows the spectacular glacier-carved valley of the river Zêzere that runs northwards. After 2km, stop at Covão da Ametade and explore on foot. Birds here include Dipper, Western Bonelli's Warbler and Firecrest, along with other tree-dwelling birds.

Going down the river valley towards Manteigas, it is worth stopping and exploring. Along the river Dipper can be seen. At Manteigas, there is a colony of Pallid Swifts (look for them near the petrol station or around the church).

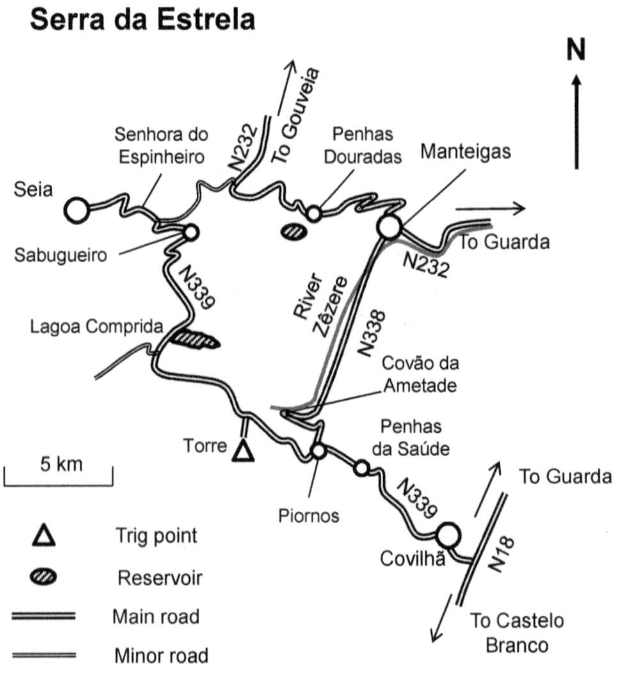

From Manteigas, it is worth driving uphill again on the road to Gouveia which will take you to Penhas Douradas, an area covered with Scots Pine plantations, where it is possible to find Western Bonelli's Warbler and Firecrest – the surrounding open fields have Woodlark and Tawny Pipit. A stop at the pousada usually provides Red-rumped Swallow and sometimes a fly-by Montagu's Harrier.

b) Western side

The other main road into the serra starts at Seia. If driving up from here, it is suggested that one makes an initial stop at Senhora do Espinheiro (near an isolated petrol station) and take a look around – this is a good spot for Ortolan Bunting in May and June. A few kilometres ahead one arrives at Sabugueiro, a village located at 1000m height. By driving through the village, as the road slopes downwards, one finally reaches the bridge over the river Alva.

The road proceeds upwards and after a few kilometres a large dam wall appears on the left – this is Lagoa Comprida ('long lake'). It is possible to park immediately before or after the dam wall and explore on foot. This is another good place for Ortolan Bunting; look out for Rock Thrushes, which have been recorded here in spring.

After going on for a few more kilometres, forever winding uphill, the road finally enters the treeless top plateau at about 1800m height. This is now Skylark, Tawny Pipit and Wheatear territory – all these species are fairly easy to spot from the road.

Turning right at the next crossroad will take you to the very top at Torre; while enjoying a cup of coffee at the highest point of mainland Portugal, it is possible to see Black Redstart in summer. A walk around should provide Skylark and Wheatear and sometimes the odd raptor. In winter there are usually a few Alpine Accentors and sometimes Water Pipits or Snow Buntings, but it is important to bear in mind that in this season the place can be overcrowded and it may be impossible to park, especially on weekends.

After leaving Torre, drive down to the next roundabout and turn right towards Manteigas. A huge rock appears on the left – it is called Cântaro Magro and is definitely worth a stop. In spring this is a prime spot for Rock Thrush which often sings close to the road. In winter this is a regular haunt of Ring Ouzel and Alpine Accentor.

By driving along this route you will eventually reach Piornos, where this road meets the 'eastern' one, described above.

Serra da Gardunha

Geographical coordinates: 40°04'49.20"N, 7°31'34.70"W

Just a few kilometres south of serra da Estrela, in the Beira Baixa province, lies serra da Gardunha, a smaller and lesser known mountainous area where it is also possible to find many higher altitude species that are uncommon in most of the country. The landscape is dramatic with large granitic rocky outcrops and huge rounded rock blocks scattered all over, forming a chaotic, beautiful backdrop. Severe fires hit the serra in recent years and destroyed considerable areas of woods, mainly pine. Vegetation is slowly recovering and large extensions are now covered with dense scrub with patches of Sweet Chestnuts and oaks. In the northern slopes there are still some large pinewoods. At lower altitudes there are vast extensions of Sweet Cherry orchards.

This is a good place for warblers and in the breeding season it is possible to find several species in the scrubby areas: Dartford, Sardinian, Subalpine and Melodious Warblers and Whitethroat all occur. In the small Sweet Chestnut and oak woods Western Bonelli's Warbler can be heard singing, while the conifer woods hold Firecrest. There are even some Cetti's Warblers. In spring, in the upper parts of Gardunha, the song of Ortolan Bunting fills the air and the birds can be seen perched on rocks marking their territories. Rock Bunting is a common resident. Other interesting species that can be seen are Booted Eagle, Short-toed Eagle, Iberian Green Woodpecker, Lesser Spotted Woodpecker, Crag Martin, Woodlark and Blue Rock Thrush. Eagle Owl probably breeds here.

The A23 motorway is the best access to this region. Leave either towards the small village of Castelo Novo or to the town of Fundão (about 50km past Castelo Branco if coming from the south).

Many dust tracks cross the serra but some are rough for a normal car. There are however two main easier options to explore. One is from the picnic ground located some 3km above Alcongosta village (close to Fundão) on the eastern side. From there a track goes up the serra to a site where there are some radio antennas and a fire watchtower. It is also possible to leave the car at the picnic area and explore on foot using the marked paths. Another option is to explore from Castelo Novo also on the eastern slope from where a dust track leads up to the summit itself (1227m high). This track can be a bit rough at the beginning but oddly it becomes tarmac after some 3km. There are also several marked paths in this area that can be walked. Castelo Novo itself deserves a stop. Besides being one of the typical villages of Beira Baixa it has a few monuments and it can hold some interesting birds. A visit to the old castle, for instance, in the right season might produce Pallid Swift, Blue Rock Thrush or Black Redstart. Golden Oriole is not uncommon in the outskirts of the village.

A third option, perhaps not as easy, is to explore the northern slope from Souto da Casa village (some 6km from Fundão). From there a rough track leads to Carvalhal, a nice picnic area lying at 800m with some small Sweet Chestnut and oak woods (search for the football pitch, where the track begins). This track crosses pinewood and scrubby areas where most of the warblers mentioned above can be seen. Besides these species Iberian Green Woodpecker is not uncommon, nor are Woodlark and Rock Bunting. Apart from birds, Gardunha's northern slope is the only place in the world to find the endemic Liliacea flower (*Asphodelus bento-rainhae*) that can be seen flowering in spring.

International Tagus

South-east of Castelo Branco, the river Tagus forms a natural border between Portugal and Spain for about 40km, hence the word 'international', and is an excellent site for large soaring birds, such as storks, vultures and eagles. The whole region is isolated and relatively unspoilt, offering the birdwatcher the chance of seeing some of the rarest and most threatened species of the entire Iberian avifauna.

Location

Geographical coordinates (Rosmaninhal): 39°44′02.80″N, 7°08′00.00″W

This site lies approximately mid-way down Portugal's eastern border with Spain, to the east and south of the town of Castelo Branco. Full details of the area can be found on the following maps: 1:100,000: 25 and 29; 1:50,000: 25-B, 25-D and 29-A; 1:25,000: 282, 283, 293, 294, 295, 304, 305, 306, 306A, 315, 315A and 315B.

Description

This site stretches from the point where the river Erges joins the Tagus in the east until the confluence of the river Sever, downstream to the west. The soil is schistose, which lends itself to numerous natural unvegetated rocky bluffs along the banks. These bluffs are not very high; nevertheless various rock-nesting birds breed here.

The river banks and those of its tributaries are largely covered with thick bushes of a Mediterranean garrigue type, where Sun Rose species dominate. The area on both sides of the river is covered with vast expanses of Cork and Holm Oaks and some eucalyptus, as well as large open fields, some of which are cultivated. The human population density is quite low and some big hunting reserves have been created in the last two decades – most of these are fenced and access is restricted.

Birds

The International Tagus area holds one of the best selections of breeding landbirds in mainland Portugal. Above all, it is very important for raptors and other soaring birds. Resident species include Black-winged Kite, Griffon and Cinereous Vultures, Bonelli's, Golden and Spanish Imperial Eagles. In summer, there are, in addition, Black Stork, Black Kite, Egyptian Vulture, Booted and Short-toed Eagles and Montagu's Harrier, while Merlin has been seen in winter.

The rocky river valley is also important for other species, including Eagle Owl, Crag Martin, Red-rumped Swallow, Blue Rock Thrush and Golden Oriole. Black Wheatear used to occur in a few places along the river but is now very rare and may soon disappear from the area.

Black Stork

The surroundings consist of open areas and woods where it is possible to find a vast array of small birds. In the more open places Thekla and Short-toed Larks are abundant and Black-eared Wheatear is not uncommon, while in the wooded areas, Azure-winged Magpies are very abundant. Also common here are Red-necked Nightjar, Hoopoe, Western Orphean Warbler, Woodchat Shrike and Rock Sparrow. Other breeding birds include Great Spotted Cuckoo, Bee-eater, Calandra Lark, Melodious Warbler, Raven, Spotless Starling and Rock Bunting.

This region is the only one in Portugal where there are regular sightings of Pin-tailed Sandgrouse. This is also one of the best sites in Portugal for Spanish Sparrow, which is abundant, both in villages and in woods, and can outnumber House Sparrows in some places; it often breeds under White Stork nests or on telegraph poles.

Accommodation

Nature and hunting tourism are now starting to develop in the area and there are a few local places to stay (provided you don't mind sharing the space with hunters during the hunting season!). Apart from such local accommodation, places to stay can be found either in Idanha-a-Nova or in Monfortinho, a tourist resort famed for its thermal springs which lies a bit further north along the N240. Castelo Branco, a major town about 40km to the west, has a wider choice of places to stay.

Strategy

Spring is by far the most suitable birding season, because the variety of species is greater, songbirds are more active and the weather is usually mild. In midsummer, the temperature can rise to over 40°C, making it unpleasantly hot at midday. Though one day is enough to see most birds, the beauty and richness of the area makes it worth a stay of several days.

This is a vast area with a lot to see and a car is essential to explore it thoroughly. However, access to some places may be difficult, so the focus will be on the most accessible sites.

a) Tagus valley

The best way to approach this area is through Rosmaninhal, a small village some 50km eastwards of Castelo Branco. This village can easily be reached from Castelo Branco on the N240, turning right to the N353 where signposted 'Rosmaninhal' and driving for about 13km.

From there a few dust tracks lead to the river valley. The village is also the starting and ending point of a long pedestrian route called 'GR29 – Rota dos Veados' (the 'deer route'). Using this kind of marked open track is probably the best option for those wishing to walk. As elsewhere there are private properties where access is not possible.

The area around the village consists mainly of open fields and Holm Oaks and holds a variety of breeding birds, including Red-necked Nightjar, and Short-toed and Calandra Larks. This is also a good area for Western Orphean Warbler, although patience is required to see them – they are usually located by the song. Pin-tailed Sandgrouse are known to occur regularly around here, however the birds are very scarce and sightings are not guaranteed. Spanish Imperial Eagle can sometimes be seen.

An interesting spot lies some 14km south-west of Rosmaninhal, around the old Guarda Fiscal (border police) post of Alares. This is one of the most accessible places in the area from where to watch the river valley (it is also included in the GR29 route). There is a bird viewing point nearby ('observatório dos Alares') which allows good views over the valley and from where it is possible to see a good diversity of soaring birds.

Alares can be reached from Rosmaninhal following the road westwards to 'Soalheiras' that passes by the small Santa Madalena Chapel (if you get lost inside the village just follow the signs to the chapel). After about 5.5km from Rosmaninhal turn left onto a large unsurfaced track (not signposted). When this track forks after 2km proceed left (downwards) and after about 1.5km, the signs for GR29 are reached. Then follow on the left hand side towards Alares. A further 2km ahead follow signs to GR29.2 'Observatório dos Alares'. From there the road gets worse and it is advisable to proceed on foot. This is another good area for Western Orphean Warbler.

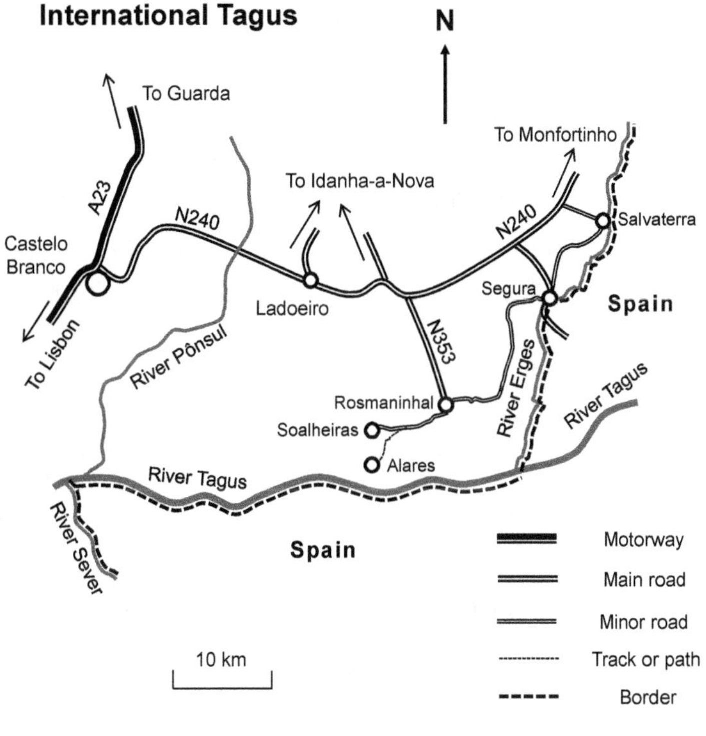

b) Erges valley

The Erges river is a tributary that runs southwards, forming the border with Spain. Its valley has large stretches of deep rocky gorges, which are interesting for their cliff-nesting birds.

Although some tracts of land are inaccessible, it is possible to approach the valley through the small villages of Segura and Salvaterra do Extremo. Segura (which lies about 17km north-east of Rosmaninhal) can be easily reached from Rosmaninhal by following a narrow tarmac road (just follow the sign to 'Segura' at the end of the N353) or from the N240. Drive past the village and proceed on the road towards the Spanish border, parking before the small international bridge over the

Erges. Here the river passes through a deep narrow canyon and it is worth spending some time at this place, as Griffon and Egyptian Vultures are easily seen. Black Storks are frequently around.

The area can also be explored by using the marked paths. From the village centre follow the signs to 'São Sebastião', 'Cemitério' and 'Rio Zona Turística'. These will lead you to the river bank.

More or less 10km to the north of Segura, at the eastern tip of this region, the small village of Salvaterra do Extremo should be visited. The river Erges flows just past the village, forming a rocky canyon where many soaring birds breed. This area can be reached from Segura via a narrow tarmac road (follow the sign to 'Salvaterra do Extremo' just past Segura) or from the N240. On arriving at the village, leave the car by the church square and walk following the signs to 'Cemitério', 'Miradouro' and 'Percurso Pedestre'. This foot track will take you through a very nice rural area where it is possible to find a vast array of small birds, and it will end at a viewing point over the river where Griffon and Egyptian Vultures, Bonelli's Eagle and Black Stork can all be seen.

Portas de Ródão

The Portas de Ródão ('Doors' of Ródão) consists of an imposing geological formation which results from the meeting of the river Tagus with the serra das Talhadas, a long and narrow quartzitic range that stretches for nearly 30km north and south of the river. This is a very good place to see a wide selection of rock nesting birds including some large soaring birds which are not common anywhere in the country.

Location

Geographical coordinates: 39°38'52.00"N, 7°40'37.60"W

The site lies near the small town of Vila Velha de Ródão, close to the Spanish border, not far from where the river Tagus enters Portugal. Detailed information about this area can be found on the maps 1:100,000: 28, 1:50,000: 28-B, 1:25,000: 314.

Description

The Tagus separates two provinces here. To the south is the Alto Alentejo, while to the north lies Beira Baixa. Although separated just by a few hundred metres, the landscape changes dramatically from one bank to the other and the same happens with the birds.

The so-called Portas de Ródão are two impressive rocky outcrops located on both sides of the river, just in front of each other. In the past, man-made terraces were built around them to cultivate Olive trees. With rural abandonment and with modern agricultural practices, most of these terraces have become covered with scrubland, consisting mainly of *Genista* and Sun Rose. Maritime Pine and eucalyptus have also been planted in the surroundings and in most of the serra das Talhadas. Along the river there are some areas of riverine vegetation.

There is a railway line that runs on the northern bank with a tunnel going through the rocks. One of the main features of this place is the malodorous paper factory located at the nearby town of Vila Velha de Ródão. Although it does not have major impact on birds or birding its odour lingers wherever you are.

Birds

This site is especially noteworthy for rock-breeding species. Black Stork, Griffon Vulture, Egyptian Vulture, Bonelli's Eagle, Kestrel, Eagle Owl, Crag Martin, Blue Rock Thrush and Raven are all present here; some breed in the Portas proper, others in nearby rocky outcrops along the serra das Talhadas. In the last few years Rueppell's Vulture has become regular and there is often one bird around, making this place one of the most reliable sites in Portugal to find this species. White-rumped Swift, a species which might be expanding in the country, has bred and is always a possibility in late spring and summer. In winter the area is often visited by Alpine Accentors. This is a good site for raptors and apart from the ones already mentioned, Short-toed and Booted Eagles, Black Kite, Hobby and Sparrowhawk can be seen regularly, while Cinereous Vulture can sometimes fly over, mainly outside the breeding season.

In the scrubby areas it is possible to find warblers like Dartford and Sardinian. Rock Bunting is not uncommon on the rocky slopes and Red-rumped Swallow is a common sight. Along the river banks Bee-eater, Golden Oriole, Nightingale, and Cetti's and Melodious Warblers are not uncommon.

Accommodation

There are not many options for staying at Vila Velha de Ródão. The best one is the Complexo Turístico Portas de Ródão in the town itself. Castelo Branco, which is a larger city about 30km to the north-east is another option as it has more to offer accommodation-wise (and which is a convenient base to explore the International Tagus as well).

Strategy

Spring is the best time to visit this area. The diversity of soaring birds is larger and the weather is usually pleasant. Summer can be unpleasantly hot and bird activity is much reduced.

This is not a big area and exploration is not very difficult. One of the best options is to go to the old medieval tower at the top of the rock on the north bank of the river and watch from there. This is a good place to see raptors and all it takes is some patience. The landscape is wonderful and vultures are usually around, sometimes flying close by as they use the valley beneath to attain height. To get there from the N18, follow the sign to 'Vila Ruivas' close to the petrol station just before crossing the river Tagus bridge. Cross the railway line and climb up the serra on the M3-13 for about 3km until a turn to the left that will lead initially to a chapel with a car park. From there it is just a short walk up to the tower.

The tiny village of Vila Ruivas is located on the western slope of serra das Talhadas on the north side of the river. It can be reached following the M3-13 (just 1.5km after the turn to the tower). From there it is possible to take a rough track that runs parallel to the serra and ends at the river bank at a place called Fonte das Virtudes close to the railway line (do not attempt to walk along the line as it is forbidden and very dangerous). The rough track is drivable with caution and it runs through Olive groves and market gardens where a good array of small birds can be seen. Walking is also an option although the way back can prove painful on the feet as the slope is steep.

The best access to the Portas de Ródão is via the A23 motorway. If you are driving from Lisbon it will take 2 to 2½ hours to reach it. Take the A1 northwards and leave after 95km where signs read 'Torres Novas'. Proceed on the A23 eastwards, leaving it after about 85km where signposts read 'Vila Velha de Ródão'. For those without a car, this place can also be reached by train (see the 'Useful Contacts and Websites' section at the end for details about train timetables).

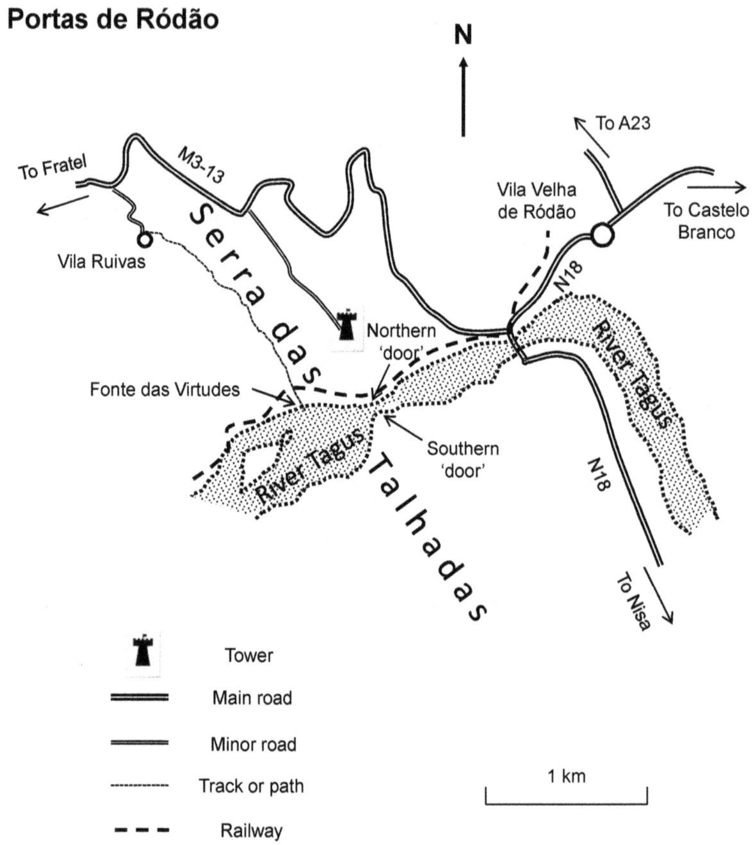

Lagoa de Óbidos

Geographical coordinates: 39°23'53.30"N, 9°12'03.80"W

This is the largest coastal lagoon in central Portugal, about 10km west of Caldas da Rainha and 70km north of Lisbon. The area is not protected and there is intense pressure both from tourism (mainly in summer) and from pollution.

The lagoon itself is interesting for a variety of waterbirds, namely ducks, waders, gulls. Greater Flamingos often turn up here. The nearby fields hold Iberian Yellow Wagtail, as well as Fan-tailed and Sardinian Warblers. Black-winged Kite and Booted Eagle can sometimes be seen there. In the surrounding hills there are some pine plantations, which hold a variety of passerines, including Crested Tit, Short-toed Treecreeper, Spotless Starling, Serin and Cirl Bunting.

Close to Vau village there are a few reedbeds, where Purple Heron, Marsh Harrier and Great Reed Warbler can be seen in spring, while Cetti's Warbler is present year round.

From Lisbon, follow the A8 motorway northwards until you reach km 70. From here there are two possibilities, visiting either the southern or the northern side.

To explore the southern side, it is best to leave the motorway at exit No. 13 and take the road to Peniche (N114), turning right where signposted Vau – this will lead you past a small village to the fields near the lagoon, close to a local airstrip. Another alternative is to leave at exit No. 15 to Óbidos and then take the small road that runs through the village of Arelho, following the signs to the airstrip. A 1km walk from here along the airfield will take you to the mouth of the river Real, at the eastern tip of the lagoon, where an observation tower has been built. This is the best place to watch waterfowl and shorebirds, especially in autumn and winter. The reedbeds lie about 500m south-west of the car park, close to the road to Vau. From Vau it is also possible to drive westwards to Bom Sucesso, close to the beach, where gull flocks are regular, especially in winter. Kentish Plover can be seen here.

To reach the northern side of the lagoon, the easiest way is to leave the A8 at exit No. 18 (near Caldas da Rainha) and look for the road that leads westwards to Foz do Arelho (N360). This village, which is reached after about 8km, lies close to the beach. From here a minor road runs eastwards along the shore and leads to a corner called Barrosa, which is also good for waders, herons, coots and sometimes Greater Flamingos. Another way to get here is as follows: leave Caldas da Rainha westwards towards Foz do Arelho; about 1km after crossing the motorway A8, turn left on Rua Alto dos Moinhos and proceed along this road; after 1.5km it turns into an unsurfaced track, which leads to the Barrosa area on the northern shore. Here it is possible to park and walk.

Paul de Tornada

Geographical coordinates: 39°26′51.80″N, 9°07′57.50″W

This small wetland is located close to the village of Tornada, about 5km to the north of Caldas da Rainha, one hour drive from Lisbon. The region is densely inhabited and land-use is intensive. The marsh, which was declared a local nature reserve in 2009, consists mainly of reedbeds, with some areas of open water here and there; the water levels change, according to the rainfall. Eutrophication is a problem.

The site is particularly interesting for birds of reedbeds. Breeding species include Purple Heron, Little Bittern and Cetti's, Savi's, Reed and Great Reed Warblers. Purple Swamphen has bred here in recent years. In winter, some wildfowl occur. Bluethroat and Sedge Warbler have been recorded regularly on autumn migration.

Birds of prey can also be seen here. Apart from the more usual Buzzard and Kestrel, regular visitors include Black and Black-winged Kites and Marsh Harrier. Goshawk breeds nearby.

In the surroundings, which hold a mix of open areas and small pine plantations, it is possible to find a variety of small landbirds, including several species typical of southern Europe, like Fan-tailed and Sardinian Warblers, Spotless Starling and Serin. Pallid Swifts breed in the nearby towns of Caldas da Rainha and São Martinho do Porto and can often be seen flying over the marsh, while they look for aerial prey.

To get here from Lisbon, take the A8 motorway northwards, leaving it at exit No. 20 (km 86), following the signs to 'Tornada'. On driving through this village, turn left, opposite a road signposted 'Salir de Matos', onto a small road (Rua do Paul) that passes close to the church at this village. Following the signs to 'Paul', after 200m there is a parking site, from where it is possible to walk to the nearby marsh. A local NGO called PATO (the name is Portuguese for duck but in this case comes from **Pa**ul de **To**rnada) has its headquarters in the yellow building next to the car park and it is possible to get information about the birds and the trails.

Peniche and Berlengas

The seaside town of Peniche is an important fishing harbour and a tourist place well known now for its almost perfect waves for surfing. It is also an interesting place to birdwatch, and the seabirds are especially noteworthy. Very close to Peniche, the small Berlengas archipelago is a place of great beauty with much wildlife of interest.

Location

Geographical coordinates (Cabo Carvoeiro): 39°21′35.70″N, 9°24′29.20″W

Peniche is some 100km north of Lisbon and the Berlengas archipelago lies nearby, about 10km off the coast. Full details of the area can be found on the following maps: 1:100,000: 26; 1:50,000: 26-C; 1:25,000: 337.

Description

The rocky coast just by Peniche is quite spectacular and the landscape is marked by the windy headland of Cabo Carvoeiro, one of the most beautiful in the country. There are vast extensions of sandy beaches and dunes around the town and ribeira de São Domingos reaches the sea here. The busy fishing harbour is one of the most important in Portugal and around the town there are large extensions of flat fields used for pasture or agriculture and small pinewoods. Close to Atouguia da Baleia, a small village to the south-east, there is a sewage treatment pool.

The Berlengas archipelago has been declared a nature reserve and consists of one larger island, called Berlenga Grande, which is surrounded by several small islets, and two further groups of islets which lie a few kilometres away to the north-west; these groups are known as Estelas (the closest one) and Farilhões, respectively. All the islands and islets are very rocky, with steep cliffs dropping almost sheer into the sea. There are no trees, and the vegetation consists mainly of small bushes and grasses, some of which are endemic to the archipelago.

Birds

This is a good area for seabirds, either breeding or on migration. The seabird populations can be found mainly on the Berlengas. Guillemot reaches its southern Western Palearctic breeding limit here, but its population has decreased precipitously from about 6,000 pairs in 1939 to a handful of pairs presently and it is now seriously endangered. The islands hold mainland Portugal's only population of Cory's Shearwater (a few hundred pairs). During the breeding season, at night, the wails of this bird make an impressive sound, echoing through the air like the calls of witches. One of the smaller islets holds the only breeding Madeiran Storm Petrels in mainland Europe. There are about 70 pairs of Shags, forming the most important site for this species in the country. Yellow-legged Gulls are ubiquitous.

In late summer and autumn, seawatching may be rewarding under favourable weather conditions. The number of Gannets passing in front of Cabo Carvoeiro can be astonishing and there usually also many Cory's and Balearic Shearwaters. All the European species of skua can be seen here and depending on the weather Storm Petrels are also a possibility. An interesting set of terns and gulls can also be seen on autumn passage, sometimes including rarities like Sabine's Gull. The passage of Razorbills, Gannets and Common Scoters can be spectacular. In winter, Leach's Petrel, Grey Phalarope and Little Auk have been seen and occasionally the odd diver turns up. The number of gulls at Peniche harbour can be very high and apart from the most common Yellow-legged and Lesser Black-backed Gulls there are sometimes rare visitors from North America, as Ring-billed, or from the Arctic, like Iceland or Glaucous Gulls.

Although the area is noted mainly for its seabirds, small parties of waders can also be seen at the streams and among them there are sometimes rarities, for instance, White-rumped Sandpiper. A few landbirds can be found easily here. Black Redstart is not uncommon in the rocky coast sectors and Peregrine Falcon also breeds. In the dunes and beaches it is possible to find Crested Larks. The nearby flat fields are used in winter by Golden Plovers and Lapwings but Dotterel and American Golden Plover have been seen there too. Cattle Egrets also use this kind of habitat, especially outside the breeding season. There are several records of Richard's Pipit.

Accommodation

Peniche is a tourist honeypot and accommodation is widely available. Note however that during the summer months prices tend to be higher and it might be hard to find a place to sleep. It is possible to stay at Berlengas too, but that should be arranged in advance. There are campsites both at Peniche and Berlengas. At Berlengas there is only a small grocery and a small restaurant. Therefore, for longer stays it is advisable to take both food and water.

Strategy

Due to its coastal location, this place can be particularly exciting during migration periods, but interesting sightings have been made at all times of the year.

Peniche can easily be reached from Lisbon in a one hour drive. Take the A8 motorway and leave it after 75km, following the IP6 and the signs to 'Peniche'.

a) Cabo Carvoeiro

Upon entering the city of Peniche, follow the signs reading 'Cabo Carvoeiro', which is reached after 3km. The best sites are just north of the lighthouse or in a secluded cleft of rocks to the left of the café, near the lighthouse. Use the stone stairs and be very careful, especially under adverse weather conditions.

Peregrine Falcon breeds nearby and sometimes hunts seabirds. It is worth taking a look at the fields close to the lighthouse, as Snow Bunting has been seen there. Just in front of the lighthouse there is a large vertical rock called Nau dos Corvos which usually holds Cormorant, Shag and sometimes Peregrine Falcon.

The rocky peninsula of Papoa, about 2km north of Cabo Carvoeiro should also be considered as it can also be a good place for seawatching and can hold some passerines, especially at migration times.

b) Peniche

The mouth of ribeira de São Domingos is worth checking. It lies just to the south of the harbour and can be reached from the roundabout before the football stadium while entering Peniche. From there, take the small road that leads to the 'Supertubos' beach then park and explore on foot. You will see the stream in front. Small parties of waders can be around and odd gulls or terns can also put on appearance. There is often disturbance here, so the best time is early morning. This area can also be approached from Molhe Leste (see below). Crested Lark is common around here.

Peniche harbour is on the south side of the town. Although birdwatching can be done from the pier itself it is best to follow the signs to 'Molhe Leste'. From there it is possible to explore on foot. This is a good place for gulls.

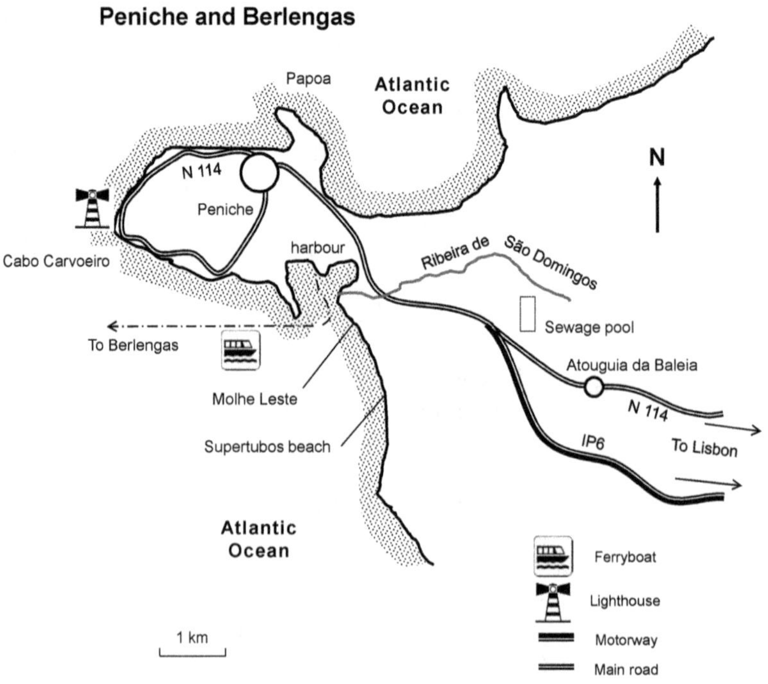

c) Atouguia da Baleia

Atouguia da Baleia is a small village some 5km south-east of Peniche. The main attraction of this place is the sewage treatment pool that attracts many waterbirds difficult to find elsewhere in the region. The pool is located at the western entrance of the village but the area is fenced off and it is not possible to enter without a permit (worth obtaining as there is a hide inside). However, it is possible to have good views walking along the western side of the fence. Usually there are some grebes, coots and ducks around. Black-necked Grebe is not uncommon in winter. This is a good place for rarities too and the long list already reported from there includes for instance Red-knobbed Coot, Ring-necked Duck, Scaup and Grey Phalarope. Richard's Pipit has been seen in the surrounding fields.

d) Berlengas

Berlenga Grande can be reached from Peniche in about 45 minutes using the regular passenger boat that operates from mid-May to mid-September. During the peak summer season (July to August) there are three daily connections, two in the morning and one in the afternoon. Unless you have time or wish to stay there for the night, the best option is to take the first one (around 09.30) and return on a late crossing (timetables and prices can be found on the website of Viamar, please refer to the 'Useful Contacts and Websites' section at the end of the book).

It should be noted that the island has a daily limit of 350 persons so it is advisable to book your trip well in advance, especially on weekends (this can be done through the web site). Outside the 'official' visitor season, it is also possible to travel to the Berlengas but that should be organised with one of the local boat companies.

Berlenga Grande is a small island (about 1km^2). Exploration is made on foot following the marked paths and it is forbidden to wander outside those. It is also possible to take a small boat tour around the island.

Yellow-legged Gulls are very numerous and can even be annoying, especially when they are nesting. Shags are also easy to spot, but the other seabirds are active mainly at night. Apart from the seabirds, the only breeding species are Peregrine Falcon and Black Redstart (numerous). Other landbirds do, however, occur on the islands, especially during migration periods, notably warblers, flycatchers, pipits, wagtails, chats and buntings. Eleonora's Falcon has occurred and several rarities have turned up here.

e) Off-shore

There are a few companies that organise fishing or leisure cruises from Peniche. It might be possible to book one of these boats for birdwatching purposes but that must be agreed in advance. The best site for pelagic viewing lies just west and north of the Farilhões, a group of islands north-west of Berlenga island itself.

The best time to try this is late summer and early autumn. Wilson's Petrels are regular from June to October. Sooty Shearwater and all four skuas and can be seen in autumn (from the middle of August onwards). Sabine's Gull is always a possibility.

Paul do Boquilobo

Geographical coordinates: 39°23'16.10"N, 8°32'34.70"W

Paul do Boquilobo is one of the most important inland wetlands in Portugal for birds. This small but attractive marsh is located quite close to the little town of Golegã, not far from the Tagus river, about 100km north-east of Lisbon. It is famous for its breeding and wintering waterbirds and particularly for its heronry. Spring is undoubtedly the best time to visit the marsh, which can, however, be interesting throughout the year (except in midsummer, when a large part of it is drained for agriculture).

The marsh is almost like a natural oasis in the heart of a heavy agricultural region. It has some areas of open water that are greatly increased in winter. The main vegetation cover is composed of willow trees. There are also some small reedbeds at a place called Braço do Cortiço. In the surroundings, apart from irrigated crops there are a few small Cork Oak woodlands.

Herons are plentiful at Paul do Boquilobo. Cattle and Little Egrets, and Night-herons are all common in the area. Squacco Heron is much rarer and, although its breeding is not confirmed every year, there are usually two or three birds around. Purple Herons also occur and some pairs of Spoonbills breed there as well. Little Bitterns used to be regular but have failed to appear in recent years.

Other breeding waterbirds have included Whiskered Tern, but this species is irregular here and numbers are highly variable, apparently depending to some degree on weather conditions. In good years, up to 120 breeding pairs have been recorded, in bad years there are none. Water Rails are resident but, as usual, very secretive. Baillon's Crake has been observed here and there is suitable breeding habitat for this species. Red-crested Pochard has bred.

Birds of prey have decreased due to habitat destruction in the surrounding areas. Even so, Black Kite occurs in summer, while Booted and Short-toed Eagles are regularly seen. Passage can bring Marsh Harrier and Hobby, while Osprey is regular in winter and during migration periods.

Among the breeding passerines, Cetti's, Reed and Great Reed Warblers are all widespread and common in the areas of aquatic vegetation while in the surroundings one can find, Bee-eater, Melodious Warbler, Woodchat Shrike, Golden Oriole and Rock Sparrow.

In winter, large numbers of ducks congregate here. Shoveler and Teal are the most abundant species, but Pintail is also quite numerous and, in some years, the marsh holds about one percent of the European wintering population.

Access to Paul do Boquilobo can be made from the town of Golegã. If travelling by car from Lisbon, it is advisable to take the A1 north and leave it after 95km, following the signs for 'Torres Novas' and then for 'Golegã', turning right after crossing the railway line. Another option, for those without a car, is to take a local train from Lisbon to Entroncamento or Tomar, leaving at Mato de Miranda (journey takes about 1 hour and 15 minutes and tickets cost €8 as of 2013) and then walk 2km northwards until the marsh is reached.

It should be noted that although quite small in size, Paul do Boquilobo is not easy to explore. The most central part is a strict nature reserve, with severe restrictions on human access. A good alternative way to explore the surrounding areas is to use the marked trail that starts at the nature reserve headquarters and circles for about 3.5km. A large part of the marsh is covered by water throughout the year, but the water level is very variable and in wet years, when rainfall is plentiful, this trail occasionally becomes inundated.

Serra de Montejunto

Geographical coordinates: 39°10′26.00″N, 9°03′06.70″W

This small serra is located about 50km north of Lisbon, rising up to 666m, and is part of the limestone chain of hills running across western Portugal. As the soil is calcareous, there are several steep vertical gorges which hold an interesting community of birds. The vegetation has been largely destroyed by fire and so most of the area consists of scrubland, with Kermes Oak the most common species. There are also small woods of pine, oak and Sweet Chestnut. At the top lie the remnants of an old convent and a few military communication antennas have been placed a few years ago. There are also some military installations nearby.

Cliff-nesting birds are the major attraction although some may not be easy to see. Species that can be found there include Kestrel (which nests in small colonies here), Eagle Owl, Alpine Swift and Raven. Bonelli's Eagle has bred in recent years. Besides these, many other species can be found. Short-toed Eagle is regular and can be seen hunting snakes on the slopes while in the scrub Dartford and Sardinian Warblers are common. Cirl Bunting is not uncommon in some areas. Hen Harrier may turn up in winter and the odd Peregrine Falcon can sometimes occur. There are several records of Alpine Accentor.

The best way to get to serra de Montejunto is to take the A1 from Lisbon, leaving it after 33km at Carregado and proceeding on the IC2 until you pass Ota, finally taking a minor road to the left, signed 'Abrigada'.

From Abrigada, the best strategy is to follow the narrow road that leads up to the summit, stopping along to way. Shortly before the top is reached, there is a turn to the left that goes along the fence of the military installations. This will lead to a small interpretation centre. The surrounding area has some nice Sweet Chestnut woods that hold a good set of forest birds. Proceeding to the very top, the area becomes treeless and an eye should be kept on the sky, as Bonelli's Eagle is sometimes around and Short-toed Eagle also turns up. This is Eagle Owl country, but as usual this species is hard to find.

Escaroupim

Geographical coordinates: 39°04′00.70″N, 8°45′26.00″W

Escaroupim is a tiny and typical Tagus riverside village some 40km upstream from Lisbon. Just in front of it, on a small island in the middle of the river, there is a superb if small heronry, which is the main birding interest in the area. Cattle Egret is the most numerous species, but Spoonbills, Night-herons, Grey Herons and Little Egrets all breed here. In 2005 at least two pairs of Glossy Ibises were found among them, in what was one of the first known breeding attempts of this species in mainland Portugal in recent times. Breeding was recorded again in 2012. This fact supports the idea that the place has the potential to become regularly used by this species in future years.

Access to the island is very difficult and in fact it should be avoided, in order to minimise disturbance. It is possible to obtain adequate and good views from the river bank, especially with a telescope. Besides herons, there are usually a few ducks around, mainly Mallards. During spring time Black Kites and White Storks are not uncommon and the odd Booted Eagle is sometimes around. The river banks are covered with dense vegetation, mainly willow trees, where it is possible to find Nightingale and Cetti's Warbler. Serin can often be seen at the car park. The area around Escaroupim is under intense agricultural use and therefore it is not very favourable for birdwatching. Even so, Black-winged Kite can sometimes be seen there; Bee-eater, Crested Lark and Fan-tailed Warbler are not uncommon.

Access to Escaroupim is made from the village of Salvaterra de Magos on the N118. From Lisbon take the A1 motorway, leaving it at Carregado and taking the A10 motorway towards Benavente. Leave at Benavente and take the N118 northwards. On reaching Salvaterra de Magos veer left at the roundabout close to the bullfight ring and follow the sign to 'Escaroupim'. After about 7km, at the entrance of Escaroupim village, take the first turn to the left (Rua João Belo) that will lead to the river bank, where it is possible to park and watch the colony.

LISBON REGION

Tagus Estuary

The river Tagus is 960km long and runs from the Sierra de Guadarrama in Spain through central Portugal, reaching the sea at Lisbon, where it forms a very wide estuary. It is the most important wetland in Portugal and one of the most important in Europe.

Its bird life is extraordinarily rich and its waterbird populations are particularly significant. Additionally, in a relatively small area, it concentrates large numbers of southern European birds, among them some of the most interesting and rare species.

Location

Geographical coordinates (Ponta da Erva): 38°50′06.00″N, 8°58′02.00″W

The estuary lies to the east and south-east of Lisbon, the capital city of Portugal, which is located two-thirds of the way down Portugal's west coast. Full details of the area can be found on the maps 1:100,000: 34; 1:50,000: 34-B and 34-D; 1:25,000: 404, 418, 432 and 442.

Description

Like many estuaries, this wetland is surrounded by densely populated and industrialised areas. Lisbon and its outskirts, together contain almost two million inhabitants, most of them living close to the river. However, the eastern part of the estuary, covering about 14,000ha from Porto Alto to Alcochete, is less densely populated and was declared a nature reserve in 1976 and a Special Protection Area (SPA) in 1994.

The reserve covers much of the intertidal mud on the left bank, where many waterbirds feed. There are also vast areas of saltmarsh, large reedbeds and lots of saltpans, many of which lie abandoned or have been converted for use as fish ponds. There are several man-made dykes along the river, which protect the surrounding plains from flood damage in wet years and from the effects of 'spring' tides, which in winter are often augmented by the strong torrent of the river. The plains here consist mainly of arable land and dry pastures, criss-crossed by a dense network of drainage canals. The banks of these canals are often heavily vegetated with a mixture of fresh water and halophytic vegetation.

Further inland, there are scattered wooded areas, where Cork Oak predominates. There are also some pine and eucalyptus plantations and, closer to water bodies, poplars, willows and densely vegetated ditches. Filled with a tangle of briars and rather moist even in summer, these are attractive not only to birds but also to butterflies, moths and small vertebrates.

The right bank of the river and some sectors of the left bank closer to Lisbon are densely urbanised and there are some heavy industry plants. Even here, despite the omnipresent, human pressure, there are places (namely Trancão river mouth on the right bank and also around Corroios, Seixal, Barreiro and Montijo on the left bank) with saltmarsh vegetation and large areas of tidal mud that hold thousands of waterbirds.

Birds

Given its geographical location and its ecological diversity, the area is important for both migrating and wintering waterbird populations. Outside the breeding season, mainly from late July through to early May, the mudflats are alive with waders. At high tide, the wader flocks

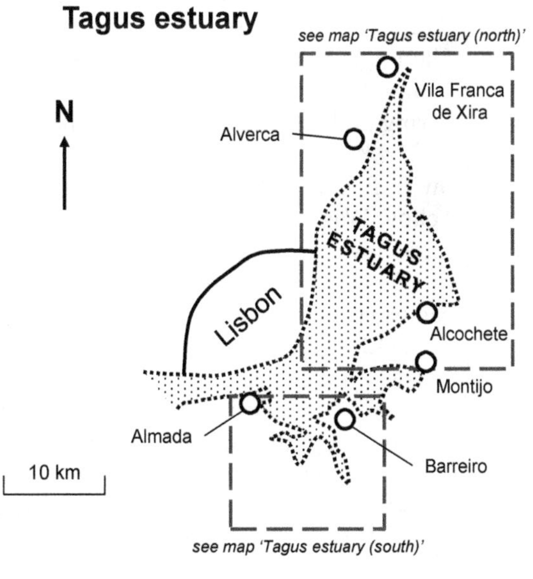

swarm across the sky, gyrating and jinking in the wake of raptors as they fly to the nontidal saltpans and inland fields.

Waders are very numerous in the estuary. Particularly noteworthy is the high percentage of Europe's winter population of Avocet which can be seen here, sometimes in vast snowstorm flocks of swirling birds. Other waders that winter in very large numbers include Grey, Ringed and Kentish Plovers, Dunlin, Redshank and Black-tailed Godwits. Most other European species of wader occur regularly on passage, often in considerable numbers: these include Curlew Sandpiper, Little Stint, Knot, Whimbrel, Bar-tailed Godwit, Greenshank, Spotted Redshank and Snipe. Other less common species, like Temminck's Stint, are regularly recorded and several rarities have put in appearances, including Pectoral and Terek's Sandpipers, Long-billed Dowitcher and Willet.

Wintering wildfowl are numerous, the most common species being Greylag Goose, Teal, Mallard, Wigeon, and Shoveler, while Garganey is less numerous and occurs mainly on passage.

Cattle and Little Egrets breed only outside the protected area but are usually widespread all over the estuary, either feeding or roosting. Greater Flamingos can be seen in good numbers throughout the year and a few thousand are often present from late summer onwards. Lesser Flamingo has been recorded on several occasions, although it is not clear if genuine vagrants are involved.

Depending upon the season, gulls and terns may be very common, forming large flocks. In late summer and autumn, it is possible to find interesting species like Mediterranean Gull (which is also present in winter) and Black Tern; rarities have included White-winged Black Tern.

During the breeding season, most waterbirds are absent and the mudflats have few birds. However, the areas around the estuary hold interesting bird communities year round and it is possible to find a substantial suite of species, many of which are typical of southern Europe.

In the reedbeds, breeding species include Purple Heron, Reed, Great Reed and Cetti's Warblers, while in winter it is possible to find Bluethroat and Penduline Tit. The rice fields are used by feeding White Stork, various herons and egrets, Glossy Ibis and sometimes Greater Flamingo. Black-winged Stilt breeds here while Gull-billed and Whiskered Terns do not breed but turn up regularly, mainly in late spring and early summer. In the marshy areas and saltpans it is possible to find Kentish Plover or Little Tern. In the agricultural and pasture areas Swift, Pallid Swift, Woodlark, Crested and Short-toed Larks, Iberian Yellow Wagtail and Fan-tailed Warbler are common while Calandra Lark, Tawny Pipit and Southern Grey Shrike can be seen in smaller numbers. There are several breeding pairs of Collared Pratincole, which can be seen flying over the fields from April through August. In winter, the fields hold Lapwing, Golden Plover and Little Bustard.

The wooded areas surrounding the estuary have a good variety of landbirds, depending on the season, which include Hoopoe, Bee-eater, Red-necked and European Nightjars, Sardinian and Melodious Warblers (the latter usually closer to water bodies), Short-toed Treecreeper, Woodchat Shrike, Spotless Starling, Azure-winged Magpie, Rock Sparrow and Serin. Less common birds include Stone-curlew, Golden

Oriole and Hawfinch.

The whole place is good for birds of prey. Black-winged Kite, Marsh Harrier and Buzzard are common residents, while at night it is possible to find Barn, Little and Tawny Owls. In summer Black Kite, Booted Eagle and Short-toed Eagle are present. Outside the breeding season, Osprey, Hen Harrier, Merlin, Peregrine Falcon and Short-eared Owl occur. Bonelli's Eagle probably breeds in the surrounding areas and is sometimes seen. There have been irregular sightings of Spanish Imperial Eagle and Spotted Eagle as well.

On autumn passage (mainly August to October), migrant species include: Great Spotted Cuckoo (until mid-August), Alpine Swift, Roller, Redstart, Whinchat, Wheatear, Whitethroat, Sedge, Subalpine and Grasshopper Warblers, and Pied and Spotted Flycatchers. Ortolan Buntings can occasionally be found in the stubble fields.

A few exotic species are to be found around the estuary. Crested Myna, an introduced species from Asia, is locally common in the more urban habitats around Corroios and Seixal, and it is now breeding. Yellow-crowned Bishops and Black-headed Weavers are not uncommon in some rice fields, while Common Waxbill is widespread.

Accommodation

The whole area lies some 30 minutes from Lisbon, which is by far the most convenient centre for exploration and where accommodation facilities are numerous and varied. It is also possible to stay at Alcochete, Almada, Montijo, Porto Alto or Vila Franca de Xira.

Strategy

The Tagus estuary is interesting at all times of the year, but in terms of waterbirds, numbers are higher outside the breeding season.

It is not an exaggeration to suggest that two or three days are necessary to cover all the habitats.

a) Ponta da Erva plains

This is a clearly delimited flat area, lying between the Tagus and its smaller tributary, the Sorraia river. The easiest way to get here from Lisbon is to leave the city northwards on the A1 towards Porto, veering right at km 25 (where signposts read 'Vila Franca de Xira' and 'Évora'), onto a metal bridge crossing the Tagus. The road here is the N10, which runs eastwards to Porto Alto and forms the northern boundary of the plains.

At the southern tip of these 'plains', called Lezíria Sul (birdwatchers often refer to it as 'Ponta da Erva', although this name actually corresponds to the very tip), the two rivers meet (point D on the map). The whole of the area is rather featureless, almost treeless and with very few houses. Most of the land is ploughed and cultivated, or left as dry pasture. In recent years, large expanses of rice fields have replaced former pasture land.

The plains are surrounded by a dyke that prevents winter flooding and they are criss-crossed by many ditches, all looking rather similar. For many years, this area used to be open and access was free, but this is no longer the case. The tracks are now blocked with gates, which are sometimes open, but it is not recommended to get in without an access card, as there is always the risk of finding them closed on the way out.

The magnetic access card, which is necessary to open the gates, can be obtained at the Associação de Beneficiários da Lezíria Grande, which lies on the N10, next to the petrol station about 4km east of Vila Franca de Xira (see map). The card service desk is not open every day, so it is advisable to contact it in advance (please see the 'Useful Contacts and Websites' section at the end of the book). The card costs €25 (as of 2013). Once you obtain the access card, you can enter and re-enter as many times as you wish, without paying any extra fee.

There are three roads into this area. All of them have gates as explained above. The first one runs to the right about 300m after crossing the Vila Franca de Xira bridge. This is the Cardal gate. After passing the gate, take the wide, unsurfaced track that leads southwards through the fields. It is passable throughout the year, but may be in bad condition after the onset of winter rains. The second gate called Giganta lies about 2km east of the first one and the track meets the previous one after a few kilometres into the plains. The third gate lies 5km east of the bridge and is called 'Ponta d'Erva'. This untarred road takes you directly through extensive rice fields and pastures to Ponta da Erva itself (point D on the map).

Whatever option you take it is advisable, when driving between the fields on the ubiquitous dust tracks, to stop frequently and search for birds. All these roads are bordered by fences that run on wooden poles, which are often used by perching birds. Take care here as mounted riders are as frequent as heavy farm machinery. The rice fields, accessible from both tracks, usually hold large numbers of waterbirds. Hundreds of White Storks, Greater Flamingos, Glossy Ibises and Black-tailed Godwits can often be seen here.

If you choose to enter through the Cardal gate, then after about 9km, the road forks (point A on the map, called 'Salina da Saragoça'). To the south (500m), there are three fish ponds, accessible but only on foot; they can hold interesting waders, ducks and, sometimes, Greater Flamingos, together with Bluethroats in autumn and winter (beware of bulls!). Proceeding eastwards from point A for 700m, take the first track to the right, near a farm entrance (point B on the map).

After about 2km, a large building appears on the left – this is the new visitor centre called EVOA (Espaço de Visitação e Observação de Aves), which was inaugurated in late 2012. It provides access to a set of three ponds and hides. A fee is payable to enter this area (€12 as of 2013). Alternatively, you can skip EVOA and proceed a few hundred metres onwards to the main dyke. Here, you can leave the car and walk along the dyke, either to the right or to the left. In summer, there are Purple Herons, Marsh Harriers, Iberian Yellow Wagtail and Great Reed Warblers to be seen. The vegetation along the dyke is good for migrating warblers and flycatchers in late summer and autumn and for Cetti's and Sardinian Warblers throughout the year, while the reedbeds have Bluethroat and Penduline Tit in winter.

Proceeding eastwards from point A but not turning right to point B, continue ahead until the road ends in a T-junction, which is point C on the map. Once here, you have reached the road that runs from Ponta d'Erva gate. From this point, turn right and take the road that leads southwards to Ponta da Erva (point D on the map). Here Greater Flamingos and wildfowl can be seen, along with waders and terns. It is important to reach this point before midday as after that time, the sun shining on the water or mud precludes the use of a telescope.

Back at point C, drive 5km northwards and just before the main road you will discover an area of extensive rice fields – in spring, these fields hold Purple Heron, Black-winged Stilt, Whiskered Tern and Collared Pratincole, while in winter there are usually many egrets, waders, gulls and sometimes Greater Flamingos.

b) Pancas

This area lies on the left bank of the river Sorraia. Its landscape differs significantly from that of Ponta da Erva, with many wooded groves of pine and Cork Oak. A large belt of open fields separates the woods from the saltmarsh.

From Porto Alto (see previous section), take the N118 to the south, following the signs to 'Setúbal', and turn right after 10km, where a small signpost reads 'Pancas' (at km 21.5). The road can then be followed and after about 1km, it forks to the left (unsurfaced road).

Proceeding straight ahead on the tarmac following signs for 'Bate-Orelhas', one finally reaches a small collection of farm buildings called Bate-Orelhas. From here, it is possible to proceed straight onwards on an unsurfaced sand track. A gate exists at the beginning of this road. It is usually open during weekdays until 17.00 and closed on weekends but as this is private property, permission to enter should be obtained in advance (do not enter even if the gate is open or you might end up being inadvertently locked in!). Here many raptor species use the thermals over the hot fields in summer. The track reaches Vasa Sacos fish ponds after about 3km (be extra careful when you enter as there are often bulls around). This place is an important tidal roost for waterbirds. However, the conversion of the saltpans into a fish farm led to an increase of water levels such that nowadays the site is not as impressive as it used to be. Even so, it is worth checking it as this place has hosted a long list of rarities for Portugal.

Back to the fork lying 1km before Bate-Orelhas, follow the unsurfaced track to the left – signposted Pancas – that leads through a Cork Oak grove. This is a good place to look for tree-dwelling passerines. It is worth frequently stopping in open areas to look for soaring raptors and Booted and Short-toed Eagles are not uncommon. At dusk, this sector is good for both species of nightjar. After about 3km this track leads to a slight elevation called Pancas. On reaching Pancas farm, the track proceeds onwards with woodland on the left (a mix of Cork Oak, Umbrella Pine and eucalyptus groves) and open fields to the right. The vista over the estuary is excellent at this point and good views of distant Greater Flamingos can be obtained from here. Black-winged Kites are often seen hunting in this area, especially towards evening. This track will lead you again to the N118 close to Alcochete after a few kilometres.

c) Alcochete

Alcochete is a small town on the south bank of the river Tagus. Here there are vast areas of the exposed river banks at low tide. Nearby there are saltpans, many of which have been abandoned or converted into fish farms. Being very close to the river Tagus itself, the wader passage can be excellent and tidal roosts often contain thousands of birds. These saltpans are best visited at the time of high tide, give or take an hour (for tide times see the 'Useful Contacts and Websites' section at the end of the book).

The whole area is excellent for birdwatching and huge numbers of ducks, herons, waders and gulls can be seen, along with flocks of Greater Flamingos (the latter are usually feeding far away, forming an impressive pink line along the estuarine skyline). Sítio das Hortas is a must for wader and wildfowl watchers and from there you have an excellent view of the estuary. Go there at low tide or mid-tide on an autumn or winter day and you will be impressed with the huge number of birds feeding on the tidal mud or roosting on the small islets of vegetation scattered about. The best time is about three hours after low tide, when all birds start to move up from the estuary. This place is good almost all year, except possibly in June and July.

For those staying in Lisbon, the area is easily accessible via the Vasco da Gama Bridge. After crossing it, leave at exit No. 3 following the signs to 'IC3', 'Porto Alto' and 'Alcochete'. Proceeding to Alcochete on IC3 you must leave at exit No. 5 signposted 'N119 – Alcochete' and proceed to a large roundabout. To reach Sítio das Hortas you should take the 3rd exit at the roundabout following the signs to 'Alcochete', 'Zona Comercial', 'Reserva Natural' and 'Sítio das Hortas'. Proceed along this road for about 2 km, past the large shopping centre and keep on following the signs to 'Alcochete' and 'Sítio das Hortas'. Shortly after passing a small petrol station on the right side, veer right onto a narrow road called Estrada das Hortas.

Sítio das Hortas itself appears on the left side after about 1km. Park the car and find a good viewing point. A small wooden platform has been built here which provides a good view over the area and there is also an interpretation centre here, although unfortunately it is not always open.

Another good place in this sector is the ribeira das Enguias saltpans complex. It is usually possible to watch Greater Flamingos here at close range, as they feed in the tanks. Flocks of Coot are often present and there are several records of Red-knobbed Coot so all birds are worth checking. Access can be made from the roundabout mentioned above. From there take the 2nd exit that leads eastwards towards a small group of houses; turn left and immediately after passing the last house, veer right onto an old and abandoned road (not signposted). This road ends after less than 1km, just before reaching a bridge in a dangerously dilapidated condition. Park before the bridge and explore on foot. It should be noted that the access to the saltpans north of the road is restricted and permission should be obtained before entering them.

To the south-west of Alcochete lies a big area area of saltpans that extends up to the village of Samouco. The construction of the long Vasco da Gama Bridge in 1998 raised a series of environmental questions, as the bridge crosses the area and several conservation measures were defined: the saltpans were fenced, closed to the public and a foundation was created to manage it in order to protect the birds and to preserve the traditional techniques of salt production. The area has now been made accessible to the public and can be visited during weekdays (weekends are reserved for group visits); an entrance fee is payable (€4 as of 2013). For more details please see the 'Useful Contacts and Websites' section at the end of the book.

The habitat here consists mainly of saltpans, many of which are still used for salt production; there are also mudflats and a sandy beach.

The saltpans are very important as a roost during high tide and are used by thousands of waders, but also by Greater Flamingos, egrets and terns.

There are two main tracks inside the fenced area that can be explored (ask for a map at the reception). Apart from the saltpans, it is also worth walking along the beach during mid tide, as it can also be rewarding (and is free of charge), but this place can be overcrowded during summer months.

Best access to the area can be made from Alcochete. Immediately before the football pitch at the entrance of the village, one must follow the signs to 'Montijo' and after a short while 'Samouco' and 'Praia' driving along the avenue that ends at a roundabout close to Forum Cultural de Alcochete (Alcochete's Cultural Centre). There one must veer left following sign to 'Samouco' taking the M501. After about 400 meters, there is a sign to the right indicating 'Salinas do Samouco Fundação para a Protecção e Gestão Ambiental'. A narrow road starts there leading to the saltpans. The first tanks to the left can be explored from the road. After about one km the gate at the entrance of the main saltpan complex is reached.

Tagus estuary (north)

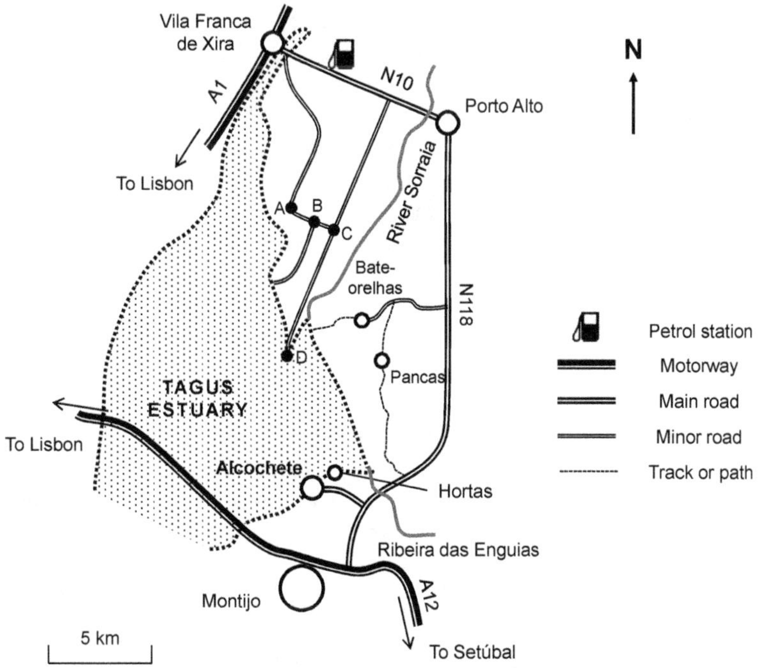

d) Coina and Mata da Machada

Between Barreiro and Coina, an area dotted with heavy industry plants and large urban expanses, there is a narrow arm of the Tagus estuary known as sapal do Coina (or sapal de Coina). Like many other places in the estuary this is largely composed of mudflats with some small reedbed patches. In administrative terms half of the area belongs to Seixal municipality and the other half to Barreiro. The latter half was declared a Regional Protected Area in 2012 and includes also a small adjacent woodland called Mata da Machada.

The area can be explored either from the small town of Coina or from Barreiro. These are connected by the N10-3 which runs parallel to the estuarine flats providing the best approach. Coina can be reached by using the A33 motorway or the N10 from Fogueteiro (if driving southwards from Lisbon on the A2 motorway, leave at exit No. 2, following signs to 'Fogueteiro'). It can also be reached by public transport, taking the train from Lisbon to Fogueteiro or the boat to Barreiro and then use bus 302 from Transportes Sul do Tejo (TST) (see the 'Useful Contacts and Websites' section at the end of the book) that runs from Fogueteiro to Baixa da Banheira, past Barreiro, stopping at Mata da Machada entrance and at Coina. A few kilometres past Coina there are some military facilities - this is the Fuzileiros school (Portuguese marines) which is a key reference point.

The estuarine area is very good for a wide array of waterbirds, especially waders. It is also a regular wintering spot for Wigeon, which is uncommon in the more urbanised sectors of the estuary. However, access to the mudflats is not very easy and the first part of the N10-3 before reaching the Fuzileiros school probably provides the best options. It should be noted that there are not many places to park and that this is a very busy road so, extra care should be taken in parking the car or crossing the road. For that reason it is best to approach the area coming from the Barreiro side, so that the river lies on the right hand side. Alternatively it is possible to access the flats from Palhais village, just before Barreiro.

The small woodland called Mata da Machada, besides being a popular place for those who enjoy picnic and jogging, also holds a good variety of woodland birds, with Azure-winged Magpies probably being the main attraction. There is an environmental education centre and several marked paths. The access to Mata da Machada lies just in front of the Fuzileiros school's main entrance and is signposted. Vehicles are not allowed, so park (do not leave valuables at sight) and walk using the marked pathway.

e) Corroios and Seixal

These areas are located outside the SPA and are not part of the nature reserve. Perhaps because of that, they have been neglected by birders for many years. Although they are surrounded by urban areas, they are very interesting for birds and access is not complicated.

The old tidal mill at Corroios is an excellent place to birdwatch. The best time is low tide when there are usually many waders around and large number of gulls often roost on the mud. Greater Flamingos and

Spoonbills are sometimes around. For those who like exotic birds this is probably one of the best places in the country to find Crested Myna. It should be noted that the mill and the surrounding space are closed on Mondays, holidays and weekend mornings.

Corroios can be reached from Lisbon by crossing the 25 de Abril bridge and taking exit No. 1, immediately after crossing it towards 'Almada'. Then, follow the signs to 'Corroios'. The access is along a large avenue (as of 2013 it is unfinished), ending after about 4km at a no-exit bridge. Once at this point the road descends towards a roundabout under the bridge, from here proceed to 'Miratejo' and 'Moinho de Maré'. Alternatively, access to the tidal mill can be made from the roundabout at N10 in Corroios (there is a BP petrol station and a hamburger restaurant there). From that roundabout follow the signs to 'Miratejo', 'Quinta do Brasileiro', 'Quinta do Rouxinol' and 'Almada' reaching the roundabout mentioned above. This area is also accessible by train either from Lisbon or Setúbal (Corroios station – see the 'Useful Contacts and Websites' section at the end of the book).

From Miratejo it is possible to reach the sandy peninsula of Ponta dos Corvos. The views over Lisbon are outstanding and birdwatching is reasonable. There are mudflats areas and sandy islands in the river which are worth checking as they are used as a high tide roost by some waders like Kentish Plover, and large numbers of gulls. Some old and dead trees nearby hold a Cormorant roost and are regularly used by Osprey. Black-winged Kite may also be around and the flat fields are good for larks and pipits. To get there proceed as if you were going to the mill (as described above). Some 300m after passing it, veer right following a signpost reading 'Bombeiros'. Proceed to the supermarket (ignoring then the sign to 'Bombeiros') until you reach a roundabout. There follow the signs to 'Escola básica de Miratejo' and 'Ponta dos Corvos'. The access track (which is usually drivable) runs along a sandbar very close to the mudflats of sapal de Corroios and after a while it begins to run parallel to an extensive sandy beach beside the river Tagus itself.

The Seixal area is also good for birdwatching. The bay is worth a look as there are usually a few gulls around, including Ring-billed Gull in winter. There are some old tidal mills with some high tide wader roosts nearby. The Paulistas mill lies close to the Benfica football centre and can be reached from the ferryboat station of Seixal following the river bank to Instituto Hidrográfico and veering right onto a mud track immediately after a football pitch by the main road. This town can be reached from the N10 via Corroios – proceed southward, past two roundabouts and then veer left to Seixal. It can also be reached from Fogueteiro (see previous section on how to get there) taking the N10 but in a northwards direction and then following signs to 'Seixal'.

f) Alverca and Forte da Casa saltpans

Over the years, the north bank of the Tagus river has suffered from heavy urban and industrial development. From the ornithological point of view, only a few interesting places have been spared. Alverca and Forte da Casa saltpans, taken together, is one of them.

The landscape of this area is composed by old saltpans, dry pastures and mudflats on the river shore. Until recently this site was under threat although it was declared an Important Bird Area (IBA). More recently, a leisure park has been created, extending from Alverca south to Póvoa de Santa Iria and a network of trails has been marked, making access easier for visitors.

This is a good place to see some open country species and also waterbirds. Red-crested Pochard has been breeding there regularly and this one of the few places in the estuary where it does so. Some of the tanks are used as tidal roost for many waders and besides the more common species, major rarities like Terek Sandpiper have been recorded here. In the drier and more open areas, Collared Pratincoles are regularly seen in spring and these are also the hunting grounds of a few raptor species, particularly Marsh Harrier.

The area lies about 15km north of Lisbon and is easily accessible from this city, using the A1 motorway and leaving it at the toll following the signs to 'N10' and 'Alverca'. After about 1.2 km, at the roundabout (with some colourful poles upon it) close to a large supermarket, there are two main options. The first is to follow the signs to 'OGMA' and 'Museu do Ar', crossing the bridge over the railway line. This will take you to a rough track (not signposted) at the end of the bridge that runs southwards, parallel to the railway line. This is the north tip of the complex. The track ends after a few hundred metres close to ribeira da Verdelha. From there a foot path called 'Trilho da Verdelha' will lead to Praia dos Pescadores after about 2.6 km (see below). The second option is to leave Alverca southwards on the N10 southwards, following the

Tagus estuary (south)

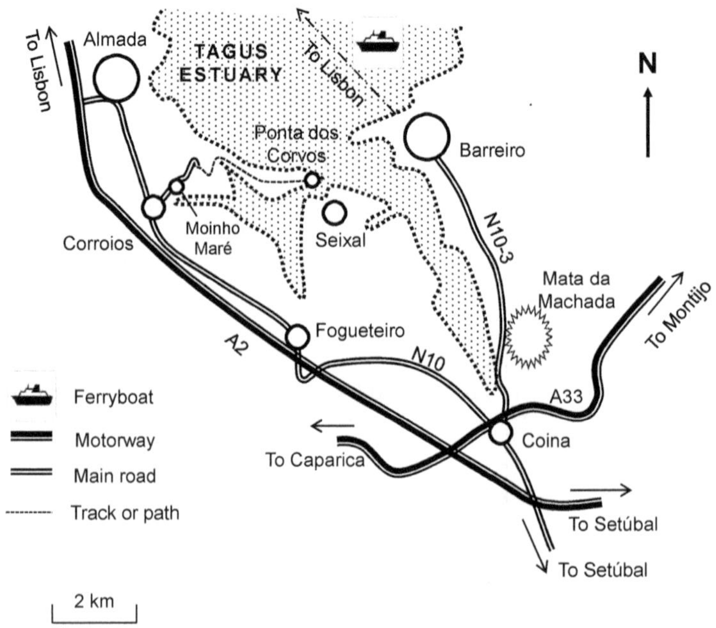

signs to 'Lisboa'. About 3.5 km later (after passing five roundabouts) one must turn left where signposted 'Parque Linear Ribeirinho', crossing the bridge over the railway line. This will take you through old abandoned factories to the Praia dos Pescadores, about 1 km ahead. There are parking facilities, a café by the river and a small interpretation center. This is the starting point of several foot paths that will allow exploration of the area.

Paul da Barroca

Geographical coordinates: 38°43'57.70"N, 8°53'38.00"W

Paul da Barroca (or Barroca d'Alva as it is also known) is a small wetland lying to the south-east of Alcochete, some 20km east of Lisbon. It has emerged in recent years as a birding hotspot mostly because of the exotic bird populations that have colonized the area. However, its ornithological interest is much wider than that; the list of species that can be found here is long and includes some specialities.

This is a marshy area that occupies the final stretch of the valley of the ribeira das Enguias, a small tributary of the Tagus. Most of the land is flat and is filled with rice fields. There are also pastures and some tanks used as fish farms. Several ditches, covered with dense vegetation, mainly reeds, cross the site. In the surroundings there are a few extensive Cork Oak woods.

Paul da Barroca is interesting all year round although winter and the passage periods might be more rewarding in terms of waterbirds which are the most noteworthy species that can be seen here. In summer, Little Bittern in not unusual in some ditches while Purple Herons can be seen feeding in wetter zones, and Cetti's, Reed and Great Reed Warblers are a common presence in the reedbeds. Outside the breeding season, large flocks of Black-tailed Godwits and Glossy Ibises feed on the flooded fields. In late summer there can be many waders in the fields, especially if these are wet. Wood Sandpiper is regular by then. In winter Lapwings are common and a few Water Pipits also occur. Great White Egrets are quite regular and in recent years, Squacco Heron has turned up frequently with a few birds present almost all year round, while Night-heron is an occasional visitor to the area. Large numbers of White Storks use this place both for feeding and for breeding and many of the electricity poles are topped with nests. The water tanks can hold many ducks in late summer and winter and there are often Greater Flamingos and Spoonbills around. Rarities seen here included Marbled Teal. Purple Swamphen has also been recorded, but apparently this species is not regular here.

The pasture fields to the west of the road that crosses the Barroca valley can hold small flocks of Little Bustards.

Paul da Barroca is also good for raptors: Booted and Short-toed Eagles are not uncommon and there are regular sightings of Bonelli's Eagle (there might be a pair breeding nearby). This area is a feeding ground for Marsh Harriers and there are usually one or two around. Black-winged Kite is also seen regularly and uses the area for hunting.

The area can be filled with swallows feeding over the valley especially in late spring and summer and Red-rumped Swallow breeds here.

Apart from Common Waxbill, most of the exotic bird species were introduced during the 1990s. Both Yellow-crowned Bishop and Black-headed Weaver are now well established breeding species. Red Avadavat can also be seen at times, but is apparently more irregular and probably not well established yet. It should be noted that most exotic species breed from late spring onwards and that in winter weavers and bishops become duller with a sparrow-like appearance.

The surrounding woodlands are good for some landbirds like Bee-eater, Woodlark and Woodchat Shrike.

The area can be easily reached from Lisbon by crossing the river over the Vasco da Gama bridge, leaving the motorway at exit No. 3 and following the signs to 'IC3', 'Porto Alto' and 'Alcochete'. Proceeding on the IC3, leave at exit No. 5 signposted 'N119 – Alcochete' and drive ahead to a large roundabout. From here proceed on the N118 following the signs to 'Pegões' and 'Barroca d'Alva'. After 3km, another roundabout is reached, where you should veer left following the sign to 'Barroca'. You will enter a narrow road which crosses the valley and from where it is possible to explore the site. This road crosses several ditches, which are worth inspecting. The flooded fields on the right hand side of the road often attract many waterbirds. Note that the entire valley is subject to floods and in wet winters it can be partially under water.

Black-headed Weaver

Lisbon

The city of Lisbon is built upon an ancient settlement site and this fact is reflected in the varying ages of its districts. Until recently the river prevented expansion to the south, hence the city grew northwards and westwards. The city offers the birdwatcher with no more than a few hours to spend some interesting surprises.

Location

Geographical coordinates: 38°42′26.00″N, 9°08′10.80″W

Lisbon is the capital city of Portugal and is located two-thirds of the way down Portugal's west coast on the northern bank of the Tagus estuary. Full details of the area can be found on the following maps: 1:100,000: 34; 1:50,000: 34-D; 1:25,000: 417, 431.

Description

Near Lisbon's centre, between Praça do Comércio and Rossio, it is possible to find the oldest buildings (most are post 1755). The further towards the north and west areas of the city, the more recent become the buildings.

In the western half of the city, the Monsanto Park forms the only really large green area. Nowadays It is crossed by a dense road network, traffic is usually heavy and it is not advisable to wander too far into the park with expensive optics. However, the Tapada da Ajuda, which lies on the southern part of the Park is surrounded by a two metre high wall, and is relatively free of traffic. It is an interesting place, consisting of wooded and semi-open habitats.

There is an artificial sea-wall, which extends from Algés (in the west) to Moscavide (in the north-east). Further to the north-east, near Parque das Nações, the intertidal mud of the Tagus estuary become exposed at low tide. Once an industrial area, it was reconverted in late 1990s for 'Expo 98' (the 1998 Lisbon World Exposition). Now it has modern infrastructures and, most important from the birders point of view, it has a very nice walkway which ends at the river Trancão mouth in the north-eastern limit of the city which allows good views over the estuary.

Birds

Although not one of the more famed birdwatching sites in the country, over 140 species have been recorded within the city limits so far. During migration times (April-May and August to October), passage can be heavy and continuous with some of the summer species typical of Spain and Portugal passing through. In early autumn, migrants occur on passage, including chats, warblers and flycatchers.

Common breeding species in Lisbon include Swift, Pallid Swift, Black Redstart (which breeds over most of the city, except possibly in the most modern built-up areas), Fan-tailed and Sardinian Warblers, Short-toed Treecreeper and Serin, while Firecrest and Spotless Starling are present but harder to find. The Monsanto Park holds Green and Great-spotted Woodpeckers, Iberian Chiffchaff, Firecrest and Long-tailed and Coal Tits. In recent years Ring-necked Parakeets have become established and are now easy to see. Other parrots species occur, namely Blue-crowned Parakeet and Senegal Parrot. Crested Myna is slowly establishing itself and has still a very localised distribution.

On the mudflats at the north-eastern part of the city it is possible to see a large variety of waders especially in winter or during passage periods. Common species include Avocet, Black-tailed Godwit, Redshank, Dunlin, Kentish and Ringed Plovers, Turnstone or Sanderling. There are usually many herons around and Spoonbill and Greater Flamingo are not uncommon, while sometimes Marsh Harrier is seen hunting. The river just in front of Lisbon can holds interesting terns and gulls on passage.

Species which have occurred within the city limits include White and Black Storks, Booted Eagle, Black Kite, Peregrine Falcon, Alpine Swift, Subalpine Warbler, Woodchat Shrike, Golden Oriole and Hawfinch. Crossbills pass south over Lisbon in irruption years.

Accommodation

There are hosts of hotels and pensions in Lisbon. Any travel or tourist agency will be happy to arrange something and even in the high season it is usually possible to obtain accommodation.

Strategy

Lisbon can provide good birding at all times of the year, but during autumn and winter months most interesting birds are to be found along the shore of the Tagus.

A map of the city can easily be obtained at every hotel or at tourist offices. It is possible to get to most places in town using public transport.

a) Southern part

In the old quarters of Bairro Alto, Chiado, Rossio and Alfama there are important colonies of Pallid Swift, and these birds can be seen from late March to late September (Swift also breeds throughout the city, mainly in the modern parts).

The old garden of Escola Politécnica, half way between Largo do Rato and Príncipe Real (in the old part of the town) is a must. An entrance fee must be paid (€1.50 as of 2013) but it is worth the visit not only for the birds but also for its interesting flora. It is a good place to see the exotic parrots and several passerines.

During the post-breeding migration it is sometimes also worth looking for terns in the middle of the estuary. The ferryboats that connect Lisbon with the southern bank can be a good and cheap option to cruise the river. One of the best is the one from Lisbon to Barreiro and its terminus is at Praça do Comércio. Besides terns, there is always the possibility of finding Mediterranean Gull which is not uncommon from July onwards.

b) Western part

Some parts of the large Monsanto Park are too noisy to make birdwatching pleasant. Even so, using the network of trails it is possible to find quieter places where some species that do not occur inside the more urban areas can be seen. Access to this area can be made from the A5 motorway that connects Lisbon to Cascais, following the signs to Montes Claros.

Side by side with Monsanto, the Tapada da Ajuda holds several interesting breeding species, like Fan-tailed and Sardinian Warblers, Short-toed Treecreeper, Spotless Starling and Serin, as well as the exotic Ring-necked Parakeet. Outside the breeding season, other species occur, and early morning autumn migration can be heard as well as seen. The Tapada is quieter than Monsanto, as there is not so much traffic. It lies very close to the 25 de April bridge and is best reached from the Alcântara area (entrance is located near Rua Jau – you may be asked for identification at the entrance).

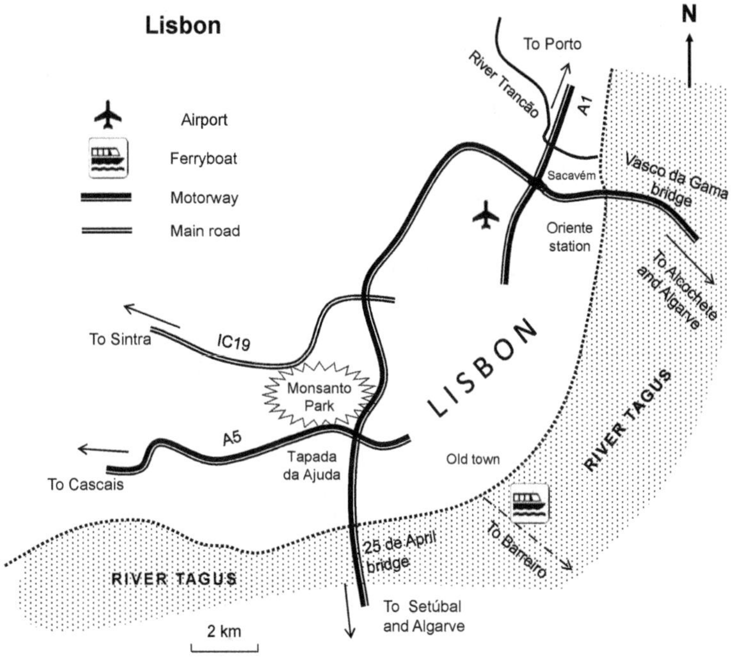

c) Eastern part

The estuary mudflats in the north-eastern part of Lisbon can be explored from the Parque das Nações area. This is easily accessible using the subway (getting out at Oriente station). From there it is possible to walk along the river until reaching the Trancão river mouth (or take a local train northwards to Sacavém). This walk should be done at low tide or mid-tide. The best place is the mouth of the Trancão itself. From there it is possible to scan the horizon and the nearby island (mouchão da Póvoa) searching for the Greater Flamingos flocks or hunting Marsh Harriers.

Costa do Estoril and Sintra

The stretch of coast between Lisbon and Cascais is known as Costa do Estoril and is famous for its beaches and expensive hotels – sometimes this coast is also referred to as Costa de Lisboa. To the north rises the romantic serra de Sintra which meets the sea at Cabo da Roca, the westernmost point in continental Europe. From the birding point of view this area offers a wide range of species from seabirds to woodland birds.

Location *Geographical coordinates (Cabo Raso): 38°42′34.00″N, 9°29′06.00″W*

This region lies just to the north and west of Lisbon. Full details of the area can be found on the following maps 1:100,000: 34; 1:50,000: 34-A, 34-C; 1:25,000: 415, 416, 429 and 430.

Description

What were once small fishing hamlets and beach-side villages such as Algés, Carcavelos and Estoril are now incorporated into an unbroken strip of development, which grows ever westwards and northwards. The hinterland is a mixture of urban areas with cultivated fields, scrubland, factories and rubble, all backed by the serra de Sintra. From Algés roughly to Carcavelos the bank of the river Tagus forms the coastline, which alternates between sandy beaches and rocky sectors.

The river meets the open sea at Carcavelos and from there onwards the habitats are much the same. Between Cascais and Guincho the coast is rocky but low. At Cabo Raso it makes a sharp turn northwards and at Guincho it opens spectacularly into a wide sandy beach but soon becomes rocky again and high northwards until Cabo da Roca is reached.

North of Cascais and along the coast lies a large pinewood known as Quinta da Marinha which connects with the nearby serra de Sintra. This small serra, a miniature mountain range really, reaches 529m and stretches east-west for about 10km, ending in the Atlantic, where it forms Cabo da Roca. The natural vegetation of Sintra has almost entirely disappeared, reduced to very small scattered patches and the serra is now largely covered with pines, eucalyptus, acacia and several other exotic plant species.

A protected area has been created covering the serra mainly and the coastline north of Guincho beach.

Birds

From August onwards the beaches form overnight roosts for gulls and terns. In misty conditions, the mouth of the estuary may fill with terns. Little, Black, Sandwich and Common Terns can all be seen, although numbers may vary greatly from year to year. Lesser Black-backed Gulls of two subspecies, and in all plumages, are numerous from August to early April and Yellow-legged Gull is seen year round. Other species of gulls occur in autumn and winter, with Mediterranean Gull numerous after gales from the west. Rarities have included Ring-billed, Laughing, Bonaparte's, Glaucous and Iceland Gulls. Interesting waders may occur outside the breeding season, on the beaches or flying over the sea during seawatches. Oystercatcher, Sanderling, Turnstone or Ringed Plover are not uncommon and this is one of the few places in Portugal where Purple Sandpiper can be found regularly.

Seabirds are one of the major attractions of this region especially from Cascais northwards. Common species includes Common Scoter, Gannet, Cory's and Balearic Shearwaters, Razorbill, Arctic and Great Skuas. The list of rarities is long and features Wilson's Petrel, Macaronesian Shearwater, Brown Booby and Sabine's Gull.

In the wooded areas of serra de Sintra, it is possible to find a good number of forest species. Short-toed Treecreeper and Firecrest are not uncommon and Goshawk, Sparrowhawk, Green and Great Spotted Woodpeckers, Crested Tit and Cirl Bunting can all be seen. Bonelli's Eagle used to breed here and may still occur.

The rocky coast holds Shag, Peregrine Falcon, Kestrel, Black Redstart

Mediterranean Gull

and Blue Rock Thrush while the coastal dunes are home to Crested Lark. On the higher and rocky zones of the serra itself, Snow Bunting and Alpine Accentor occur in some winters and Ring Ouzel has been recorded on passage.

Pallid Swifts are common during spring and summer over most of the area and probably breed in many built-up places while Crag Martin is a local winter visitor.

Azure-winged Magpies have recently moved into this area and are seen mainly in the lowland pinewoods and scrubs between Cascais and Guincho beach. Another more or less recent species is the Crested Myna. This exotic species from Asia is closely related to starlings and has been introduced in the country – this is one of the best places to find it.

Accommodation

This is a tourist-saturated area and accommodation should not be a problem. Cascais, Estoril and Sintra have a wide range of hotels to choose from, and the whole region can easily be covered from Lisbon as well.

Strategy

This site is most interesting during migration and winter time, although a few interesting breeding birds can also be found inland.

Cascais is about 25km west of Lisbon and can be easily reached from the city by using the A5 motorway or the coastal road N6. Sintra lies about 20km to the west of Lisbon and the most direct route to get there is via the IC19.

a) Algés to Cascais

This sector can be explored from the N6. The 'Marginal' as it is known, hugs the coast here and it is merely a matter of motoring along to likely beach sites. Do not stop on this road or attempt to birdwatch while driving here; use the available lay-bys which are mainly available in a west-east direction. The best strategy therefore is to drive from Cascais towards Lisbon as this allows direct access to the beaches and car parks. The area can also be explored using the train (the 'Linha'). The rail line from Lisbon to Cascais passes close to the sea and the stations are clearly marked. Large parts of the coast can be covered from walkways, so exploration on foot along some stretches is also a possibility.

All the beaches are worth examining and autumn/winter is probably the best time for visiting. Summer months should be avoided (there are few birds and too many people). At all times of the year there is much activity on the beaches, so early morning is the best time to birdwatch.

Starting at Cascais and moving towards Lisbon to take profit of best access and of the existing car parks, it is worth stopping at Parede beach. This is a regular spot for Mediterranean Gulls. About 3km eastwards lies Carcavelos beach, where there is often too much disturbance. However, early morning visits during winter have produced unusual species like Ring-billed and Laughing Gulls. The nearby fields close to the old fortress of São Julião da Barra are a good place to find the exotic Crested Myna.

Past Carcavelos, Santo Amaro de Oeiras beach is the place to look for wintering Purple Sandpipers that can be found on the rocky areas just by the coastal walkway.

Algés beach is the closest to Lisbon and is recommended for passage waders and terns, especially during easterly winds (do not ignore the mouth of Ribeira do Jamor further west, near Cruz Quebrada railway station).

Besides the above mentioned possibilities, there are a few other beaches that might be worth exploring. For instance Paço de Arcos and Caxias are sometimes good for gulls, especially in winter.

b) Cascais to Cabo Raso

Start by exploring the Cabo Raso area as seawatching is most productive around dawn. This headland lies some 6km north-west of Cascais and is the best seabird watchpoint around Lisbon. It is located just by the N247 and can be easily identified by the red metal tower of the lighthouse. It is possible to park safely and it is even possible to seawatch from the car (most useful under bad weather conditions). The best time is from late summer to winter although spring migration can also be rewarding. The period immediately after the passage of a weather depression is particularly favourable. From August to May onshore W/NW winds and rain bring Cory's and Balearic Shearwaters close inshore. The list of seabirds seen off the Cabo Raso is extensive and includes many common species as well as some vagrants. Apart from seabirds, the area around Cabo Raso also holds some small landbirds. There is a boardwalk that crosses the dunes to the north, which can be reached from the N247, about 2.5km northwards towards Guincho. It is also possible to visit this place by using the bus that departs from Cascais train station. If you use this alternative it is advisable to check the return time table (see the 'Useful Contacts and Websites' section at the end of the book).

The road from Cabo Raso to Cascais runs along the coast. There is also a bicycle lane and a walkway along most of the road. To the northern side there are some wooded and scrubby areas and some dunes. Pay attention as Azure-winged Magpie is now regular here.

On the way to Cascais, there is a tourist spot called Boca do Inferno (literally 'hell's mouth') some 2km before the town centre (easily reached on foot from there). The geological aspects are interesting but this can also be a good place to seawatch, especially for those without a car and staying at Cascais. Peregrine Falcon can sometimes be seen here and Black Redstart can be found on the cliffs.

The Cascais area itself is fairly good for birds. In the bay there are usually some gulls and sometimes there are rarities among them. In winter, seabirds like Razorbill look for shelter here. Close to the marina, there is a small palace on the sea front that in winter has a regular roost of Crag Martins.

c) Cabo da Roca

The westernmost tip of the Sintra hills runs into the sea at Cabo da Roca. This is a rugged area with high sea cliffs, some small islets by the coast and deep valleys. The cape itself is a major tourist attraction and it can be crowded, especially on weekends. Even so, it is worth walking along the fence to the south as Blue Rock Thrush can often be seen there. Autumn migration brings interesting passage species but the area is underwatched. Even so, a few rarities have been recorded like Wallcreeper, Red-breasted Flycatcher and Snow Finch and regular watching would probably reap rewards.

A quieter option is to turn right about 200m before reaching the cape, following the sign to 'Ursa' (literally meaning female bear). This is the

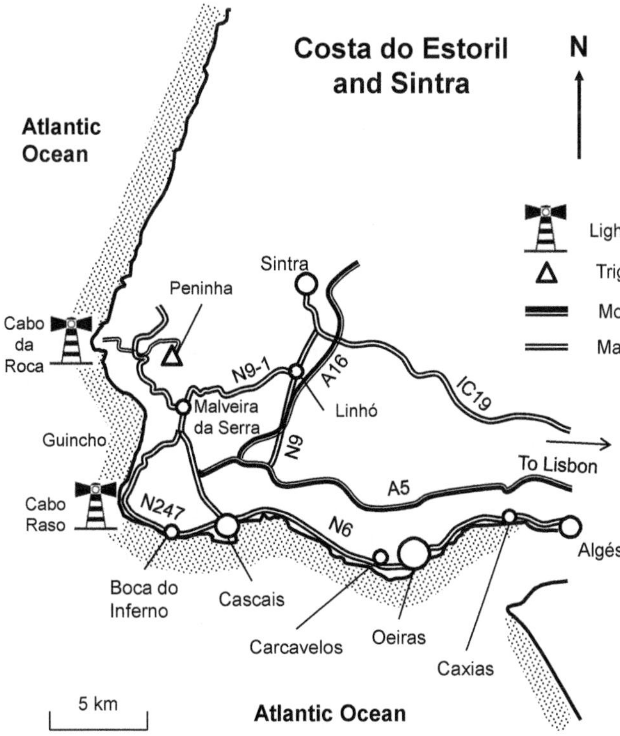

beginning of a very rough track that ends after about 600m just in front of the small Ursa islet. The valleys and scrubby areas around here can be interesting especially at migration time. If you don't have a high clearance vehicle it is wiser to park and walk (do not leave valuables in sight).

Access to Cabo da Roca is easy. If coming from Lisbon on the A5 just follow the signs to 'Cabo da Roca' at the end of the motorway, taking the N9-1 until Malveira da Serra and then the N247. If coming from Cascais, just follow the N247 westwards along the coastline, past Cabo Raso and Guincho beach. Between km 87 and 88, turn left to 'Cabo da Roca' which is reached after about 2.5km.

Cabo da Roca can also be reached using the public transport network. There are regular buses either from Cascais or Sintra train stations (see the 'Useful Contacts and Websites' section at the end of the book).

d) Sintra

Serra de Sintra with its gardens and palaces is a 'must see' for most visitors to the Lisbon region. Despite not being a prime birdwatching site, it holds some interesting species. For those who want to combine a cultural visit with some birding there are a few options to explore. For instance, the large parks and gardens like Pena or Monserrate are easily accessible (however, an entrance fee is payable and the place is often crowded, especially during weekends or summer season). One interesting and much quieter spot is Peninha, on the western part of the hills. This high vantage point, which has a small palace on the top, lies 'behind' Cabo da Roca and can be reached from the N247 close to where it veers to the cape. The views are spectacular and the place is regularly visited by Alpine Accentor in winter, while Blue Rock Thrush is resident. Raptors can sometimes be seen from here.

Another possibility to explore is to use the N9-1 that crosses the southern slope from Linhó to Malveira da Serra. Stopping along this road makes it possible to see an interesting set of woodland birds. After a few kilometres it is worth taking a look at Lagoa Azul (the 'blue lake'), a small dam among the pinewoods. This is a good location for Cirl Bunting.

For those who like trekking, there are several marked paths that can be used to explore most of this area.

Sintra can be easily reached from Lisbon using the IC19 or the A5. Both roads have heavy traffic and should be avoided during rush hours. It is also possible to get to Sintra from Lisbon by using the train. It should be noted that this is a very popular place on weekends, so for those who do not enjoy crowds it is preferable to head for there during weekdays.

Lagoa de Albufeira

Until recently, this coastal lagoon remained somehow forgotten by many birders. Nevertheless it is an interesting place for watching birds and the total number of species already recorded here is around 240. One of the main attractions of this area is the Purple Swamphens. Besides it is also a good site for breeding Purple Heron and Reed Warbler, and for some wintering specialities.

Location

Geographical coordinates: 38°31′20.60″N, 9°08′29.40″W

The lagoon lies some 25km south of Lisbon, not far from Cabo Espichel. Detailed information on the area can be found on the following maps: 1:100,000: 38; 1:50,000: 38-B; 1:25,000: 453.

Description

Lagoa de Albufeira is formed by two main bodies of water. The largest one on the west side is called Albufeira or Lagoa Grande ('big lagoon'), while the smallest one further east is called Lagoa Pequena ('small lagoon'). The former is a wide open water surface with no vegetation at all. It is separated from the sea by a chain of sand dunes and a channel is opened here every spring, in order to allow sea water to get in. The latter is also a wide open water surface but it has some vegetation along the banks and at the inner end. The ribeira da Apostiça is a tributary of Lagoa Pequena and along its final stretch forms a valley with pastures. A ditch was built at the mouth of this river creating a flooded area covered with reed and other riverine vegetation which is called Lagoa da Estacada. The area around Lagoa da Estacada and Lagoa Pequena is covered with woodland, mainly Umbrella Pines but also some mixed woodlands of oaks and pines. The surroundings of Lagoa Grande are occupied with small houses that were built after the 1974 revolution. To the north of Lagoa de Albufeira there are extensive stands of pinewoods. Along the course of the Aiana river that reaches the south side of Lagoa Pequena, it is possible to find some open fields with pastures and small market gardens.

Human activities vary according to the time of the year and like many other wetlands in southern Europe, it is under threat from a variety of sources. Pressure resulting from tourism can be heavy in summer, although people concentrate largely at the beach of Lagoa Grande. Other activities causing disturbance are motocross rides (mainly with quad bikes), all terrain treks with jeeps and fishing. Hunting is not allowed. Back in the beginning of the 1980s a fence was built around Lagoa Pequena and part of the Lagoa da Estacada. Lagoa Pequena was designated as Special Protection Area (SPA) and it is also an Important Bird Area (IBA). A visitor centre was inaugurated in early 2012 and since then it has been possible to get to the four hides that have been built here.

Birds

By the end of January hirundines start to arrive, harbingers of spring, and by mid-February the first Iberian Chiffchaffs are singing. In the beginning of March, Purple Herons arrive then Nightingales and Cuckoos sing in the woods. The noisy call of Purple Swamphens echoes through the reeds and these magnificent birds can be seen feeding using their large feet to grasp the thick-rooted waterside plants. The shy Water Rail usually hides among the vegetation and is most often detected by its call. Cetti's Warbler is quite common and Marsh Harrier breeds in the area.

Autumn migration brings small flocks of waders sometimes mixing with the last tourists at Lagoa Grande! Sanderlings are numerous at times and other species include Bar-tailed Godwit, Dunlin, Redshank and Greenshank. Terns use Lagoa Grande as a stopover point during migration or as feeding place in winter. Sandwich Tern is the most common but Common and Little Terns are also regular on passage. In July and August a careful search of gull flocks might produce Audouin's Gull. In late summer and autumn Cory's and Balearic Shearwaters often pass not far from the beach, where Crested Larks are usually around. Huge numbers of swallows and swifts can sometimes be seen feeding in the area especially over Lagoa da Estacada but also over the open fields of Aiana and Apostiça including big numbers of Red-rumped Swallows in late August and the beginning of September.

The woods that surround the south bank of Lagoa Pequena are a good place to see the resident Crested Tit, Short-toed Treecreeper, Serin and Cirl Bunting. Woodlark occurs in small numbers and small flocks of Azure-winged Magpies sometimes turn up. In the more open areas Fan-tailed Warbler is common.

This is a good area to see raptors; Buzzards are usually around. Other species that occasionally put in appearance include Black-winged Kite, Sparrowhawk, Short-toed and Booted Eagles, and even rarities like Eleonora's Falcon.

There are not many ducks in winter, but Mallard, Teal, Shoveler and Gadwall are guaranteed at Lagoa da Estacada and Ferruginous Duck has occurred here (though the origin of these birds is uncertain). Coot counts exceed sometimes 1000 and Red-knobbed Coot is a possibility. Sightings of White Storks are not uncommon and Crag Martins are regularly seen feeding over Lagoas Pequena and Estacada. Bittern has been recorded in recent years, usually among the reedbeds. The wide open waters of Lagoa Grande are the place to search for the odd diver, Great Northern being the most likely. Black-necked Grebe is not uncommon and a few birds are usually around.

Accommodation

This site is 30 minutes drive from Lisbon which offers a wide range of accommodation. The small seaside village of Sesimbra is nearby and there it is also possible to find plenty of accommodation, except maybe in summer months when hotels and pensions may be full.

Strategy

Lagoa de Albufeira can be interesting all year round.

When coming from Lisbon, take the A2 motorway southwards to Setúbal crossing the 25 de Abril bridge. After 10km leave the motorway and follow the signs to 'Sesimbra'. Proceed on the N378 through Fernão Ferro and after 6km you'll reach a big roundabout where you need to turn right to 'Alfarim', 'Lagoa de Albufeira', taking the N377. You will reach the bridge over Apostiça river 3km later. From this point onwards the road runs along the south bank of Lagoa da Estacada. Although this road is filled with heavy traffic especially on summer weekends, it allows good views over the fenced area and the reedbeds on the right side. You can use the existing lay-bys to park safely and to look for Purple Swamphens and Purple Herons, among other waterbirds.

About 2km ahead, there is a small track to the right signposted 'Observação de aves lagoa Pequena' that leads to the main gate of the fence and to the visitors support facilities. This is open on Wednesdays, Fridays and Saturdays from 09.30 to 12.30 and from 14.00 to 18.00 and a small entrance fee is payable (€2.1 in 2013). From here it is possible to have access both to Lagoa da Estacada and Lagoa Pequena. At the former there are three hides which provide a good view over the water and where Purple Swamphen is usually seen. Other birds here include herons in spring and waterfowl in winter. The Lagoa Pequena is readily visible on the left hand side from the gate and it is also possible to walk along the fence down to the bank and search the mud for waders (there is another hide which provides a view over this area).

Returning to the N377, and moving along towards Alfarim, the road crosses the valley of the Aiana river. This is an open area with some market gardens and pastureland surrounded by Umbrella Pine and mixed woods. The area can be explored using the two mud tracks that run parallel to the valley. The first one is reached 500m after the km 30 marker. It can be in bad condition and it is not advisable to drive it all the way to the end, at the bank of Lagoa Pequena (instead park safely 500m after the start). The second track starts to the right between km 31 and 32 immediately after crossing the bridge over the Aiana river. It can be traversed by vehicle all the way with some caution and it ends after 2km at the southern bank of Lagoa Grande.

The Lagoa Grande area and the beach are reached by turning right on the N377 about 4km after the visitor centre where signposted 'Lagoa de Albufeira' and 'Praia da Lagoa'. After the turn there are two options. If you want to go to the beach then you have to keep on the tarmac road following the signs to 'Mercado' (market). Proceed for 2km and then towards 'Praias' as stated before; this area can be crowded in summer, so visits to Lagoa Grande at that time of year should be avoided. In autumn and winter, water level allowing, there can be large flocks of gulls resting on the sand banks and a careful look might produce

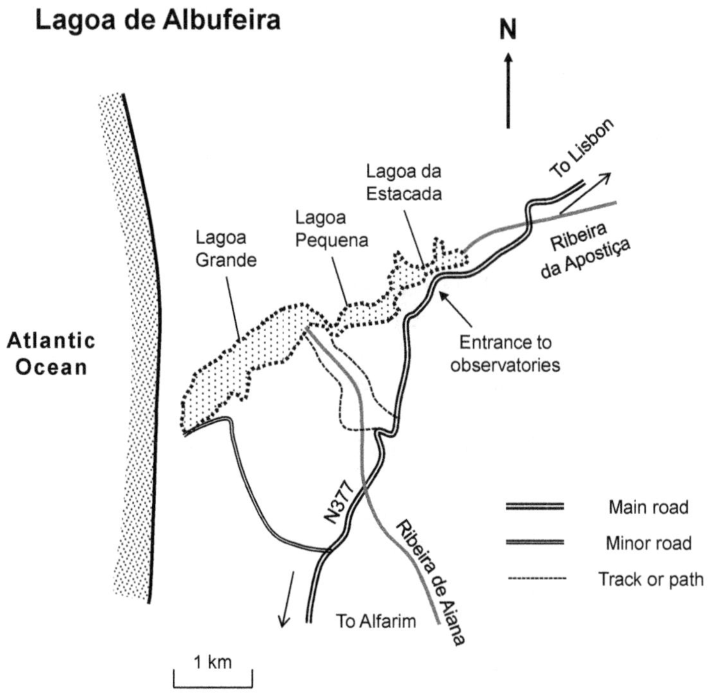

rarities. It is also a good idea to take a look at the sea and scan for seabirds. If you want to reach the middle of Lagoa Grande then you need to go straight ahead taking the unsurfaced road that will end after 2km at the bank of lagoon. In winter this location is good for divers and Black-necked Grebes.

If you are planning to visit this site in July be aware that a big music festival occurs during this month, lasting for a few days; it usually takes place in the surroundings of Lagoa Pequena making access to the whole area almost impossible. When driving on the mud tracks, extra care must be taken as enthusiastic but careless four wheel motorbike drivers race along them, especially during weekends. There have been reports of car robberies around here, so it is not advisable to leave valuables in sight if your vehicle is left unattended.

Cabo Espichel and Serra da Arrábida

Between the estuaries of the Tagus and the Sado lies an area of overdeveloped urban sprawl, some wonderful coastal scenery and an imposing limestone range, running east-west for 30km. The western end of the range reaches the sea at Cabo Espichel, an isolated promontory with small farms and a lighthouse.

This area, known as Península de Setúbal, is particularly interesting at migration times, but there are also some attractive southern European species breeding here. On the southern part of the peninsula, there are two main places worth investigating: Cabo Espichel and the serra da Arrábida, with a stop at Sesimbra between the two.

Location

Geographical coordinates (Cabo Espichel): 38°25′12.00″N, W 9°12′45.00″W

The site lies to the south of Lisbon on the west coast of Portugal. Full details of the area can be found on the following maps: 1:100,000: 38; 1:50,000: 38-B; 1:25,000: 454, 464 and 465.

Description

This is a curious area geologically, surrounded to the north-west and south by the sea and backed by the serra da Arrábida, which stretches along an axis and has superb scenery. Cabo Espichel, at the western tip, is mainly limestone karst with pockets of scrubby vegetation and some pine plantations. The soil is poor and cultivation rather desultory. The only buildings of any size standing out are the lighthouse and the large sanctuary of Nossa Senhora do Cabo.

The serra da Arrábida is not high as Iberian ranges go, but its landscape is very characteristic with limestone outcrops peeping out from a dense cover of Mediterranean scrub ('carrascais'). The vegetation is dominated by Kermes Oak, Strawberry Tree, Carob Tree and Tree Heather. At long range the inclined strata lines of limestone lend the peaks a unique aspect.

Arrábida is one of Portugal's best known natural parks, with restrictions on the construction of buildings and preservation orders on some of the older buildings. However, the serra is under pressure from two quarries, one at either end, as well as from a cement factory near Setúbal; additionally, there has been an increasing pressure from tourism.

Birds

Cabo Espichel is very dry, with a microclimate compounded by its porous limestone geology. During the breeding season the most interesting species are Peregrine Falcon, Pallid Swift, Bee-eater, Crested Lark, Black Redstart, Blue Rock Thrush and Cirl Bunting.

It is far more interesting at migration times, basically due to its strategic location on one of Iberia's flyways. The intensity of spring migration is dependent on wind, but typically will include a small regular passage of Turtle Doves, various swifts, as well as a good selection of passerines and usually peaks from late March to early May.

The autumn passage lasts from August through October to mid-November and is marked by a regular passage of trans-Saharan migrants – swifts, larks, hirundines, chats, warblers (including Melodious, Willow and Subalpine), Woodchat Shrike, flycatchers and sometimes Roller. Short-toed Larks, Tawny Pipits and Ortolan Buntings are regular. Unlike at Sagres further south, birds of prey tend not to concentrate at Espichel, but Egyptian Vulture, Eleonora's and Red-footed Falcons and Lesser Kestrel have all been observed and Short-toed and Booted Eagles are seen occasionally.

Arrábida is a picturesque area. Spring is marked by the presence of breeding birds such as Bonelli's Eagle, Alpine and Pallid Swifts, Red-necked Nightjar, Iberian Green Woodpecker, Hoopoe, Bee-eater, Woodlark and Crested Lark, Red-rumped Swallow, Blue Rock Thrush, Sardinian Warbler, Iberian Chiffchaff, Short-toed Treecreeper, Golden Oriole, Azure-winged Magpie, Serin, Hawfinch and Cirl Bunting. Falls of migrants passing southwards along Portugal's west coast or reaching the coast at daybreak often occur in these hills and in autumn a wide variety of trans-Saharan migrants can be seen.

This is also a good area for seabirds. Cory's and Balearic Shearwaters are common and the passage of terns may be interesting in season. Macaronesian Shearwaters and Wilson's Petrel have been seen just offshore and the large submarine canyon to the south of the head is worth traversing, as it creates local upwelling conditions in late summer and winter.

Accommodation

The whole area is accessible by car from Lisbon or Setúbal; both are cosmopolitan cities with a wealth of accommodation. Nearer to Cabo Espichel, the fishing town of Sesimbra has many hotels (including four-star) and pensions. There is a hotel on the road to the cape from Santana that is a convenient base from which to reach Cabo Espichel quickly in the morning to watch passage migration especially after a period of easterly winds.

Strategy

All sites can be visited in a day's trip though Cabo Espichel is definitely best in the predawn and the four hours after dawn. Arrábida is interesting during the breeding season, while Cabo Espichel is particularly rewarding during migration times.

Access to this area is not difficult. Approaching from Lisbon, cross the Tagus by following signs indicating 'A2 SUL' (the bridge toll is only paid on the northward journey – €1.60 as of 2013). About 10km southwards leave the motorway near Fogueteiro and follow the road

N378, which is signposted 'Sesimbra'. Proceed for 17km through Fernão Ferro until you reach Santana.

a) Cabo Espichel

Cabo Espichel can be reached from the village of Santana on the N378, turning right following the signs to 'Cabo Espichel'. The village of Azóia, about 10km ahead, marks the beginning of the cape proper.

The best strategy on reaching Azóia is to proceed straightforward until the end of the tarmac road. There, on the right hand side there is an enclosed garden with Fig and eucalyptus which is well worth investigating for migrating chats, warblers, flycatchers and buntings (unfortunately as of 2013 it is no longer possible to walk inside, but it is still possible to watch the place from the outside). An unsurfaced road to the left leads to the lighthouse (where the gardens are worth examining). From the lighthouse it is possible to walk down to the very tip, where two abandoned houses still exist. Seawatching is possible from here (see below). Additionally, the views over the cliffs are magnificent and there are good chances of seeing Peregrine Falcon and Blue Rock Thrush.

Local resident birds in the area include Cirl Bunting, Crested Lark and Black Redstart. The old crumbling walls of the sanctuary complex offer ready cover and should be carefully examined also. The valley between the sanctuary and the lighthouse is scrub covered and can be excellent in windy easterly conditions (which is often the case in spring) when other sites are difficult to watch.

Another possibility from Azóia is to continue for about 1km towards the cape before turning onto a rough track to the left (the sanctuary and the lighthouse are visible ahead). Leave the car here and walk through the unfenced fields along lines of Pistachio bushes to the lighthouse. From late August to early October, Ortolan Buntings are very regular here, along with Subalpine Warblers and phylloscopids including Western Bonelli's Warbler. In the early autumn, Short-toed Larks, Tawny Pipits and Yellow Wagtails are regular diurnal migrants, succeeded as the autumn progresses by Skylarks, White Wagtails and finches. Richard's Pipit and Dotterel have been recorded here. Mixed flocks of starlings are seen regularly – look for Rose-coloured, which seems to be an annual vagrant in autumn.

The north side of the headland between the sanctuary and Azóia is also worth examining and is approached via rough tracks, best travelled on foot. Not very far from Azóia, a track leads to the north opposite a line of 'half built' holiday homes, not far from a sign reading 'Chã dos Navegantes'. This track is not negotiable by car but is worth the walk – the habitat is ideal for a wide variety of migrants. Another track to the north is reached by turning north beside the old stone aqueduct. This walk is excellent for flycatchers, shrikes and buntings during migration times.

b) Sesimbra

The town of Sesimbra, half way between Cabo Espichel and Arrábida, is ideal for a stop. From Santana, just follow the signs to Sesimbra, which lies 3km away downhill. The harbour is usually full of gulls, which are worth inspecting, as rare species from the north occasionally turn up here. The odd diver has also been observed on occasions. The castle deserves a visit – not only for its far-reaching views, but also because it is an interesting place to look for passerines: Crag Martin (in winter), Black Redstart, Firecrest, Crested Tit and Hawfinch have all been seen here, along with a dozen more common species. During summer months, Pallid and Alpine Swifts are frequently seen flying over the area.

c) Serra da Arrábida

Arrábida can be approached from the east (through the town of Setúbal) or from the west (via Azeitão). The following description is for the western route. If coming from Lisbon (or from Cabo Espichel), drive to Santana as described above and at the roundabout outside the village turn through 180° until signposts read Azeitão and Arrábida. Continue along this road (N379) for about 8km, then veer right twice following the signs reading 'Arrábida'. Route 379-1 takes you into the park.

The first part goes through mixed woodland of pine and oak with some fields and vineyards in-between. Typical birds in this part include Iberian Chiffchaff, Azure-winged Magpie and Cirl Bunting. During spring nights, Red-necked Nightjar can be heard singing.

Pass the small village of Casais da Serra, to reach the more exposed heights of Arrábida, where Bonelli's Eagle, Alpine and Pallid Swifts and Azure-winged Magpie can be regularly seen.

On reaching the next cross-roads, veer right to Portinho da Arrábida. This route (379-1) also provides access to the south-facing slopes of Arrábida; this area is covered with scrub – the characteristic sclerophyllous woodland dominated by Kermes Oak. Small secluded valleys of Lusitanian Oak on the seaward slopes and in particular where the road runs down to the village of Portinho da Arrábida are well worth exploring, especially in the mornings. Iberian Chiffchaff is plentiful here and can easily be heard singing in spring. A visit to the nearby Convento da Arrábida (an old convent) is worthwhile for breathtaking views and maybe access to some of the species mentioned above. This is private property however and a fee is required to get in (more details about general conditions and timetable can be obtained at the official website, please refer to the 'Useful Contacts and Websites' section at the end of the book).

Proceeding eastwards to Setúbal, the road runs along the sea, beneath a huge rocky wall. A stop at Figueirinha beach and a careful scrutiny of the cliffs might produce Peregrine Falcon, Black Redstart and Blue Rock Thrush.

Access to most of Arrábida is open and there are many walking tracks but certain parts are restricted. It should however be noted that Arrábida is a very popular place and tends to be crowded on weekends, particularly during summertime, when the small Portinho beach may be full; at this time of year there may be some restrictions on traffic (especially on the road that connects Portinho to Setúbal). A visit during winter or spring weekdays, when the place is quieter, is definitely a better option.

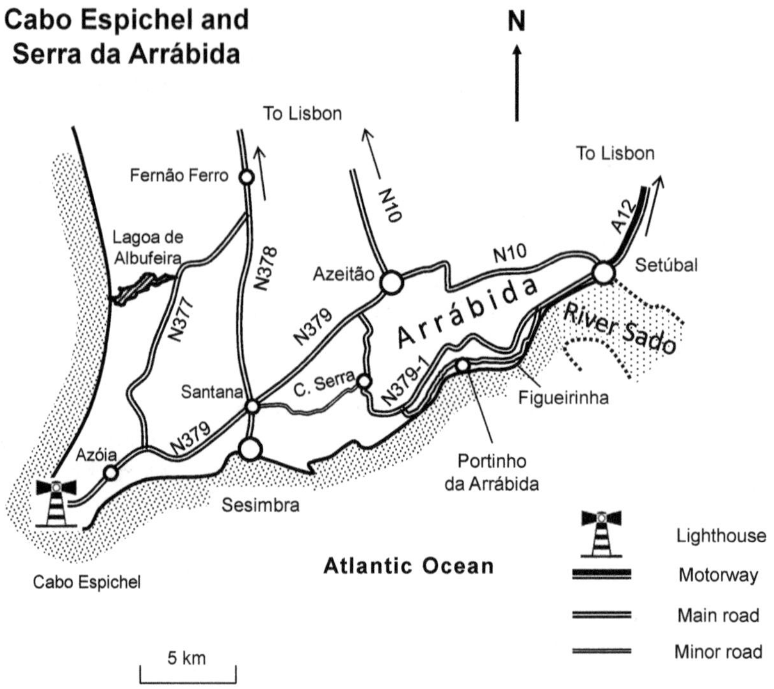

d) Seawatching

The best seawatching option in the area is Cabo Espichel. It is worth walking past the lighthouse and descending the scree slope until one reaches some old and abandoned military buildings. Although the place is too high for sustained seawatching, it is always worth taking a look, because sometimes pelagic species pass close to the shore (be careful here as the scree can be treacherous).

Another option is to hire a boat for a pelagic trip, which might be arranged in Sesimbra or Setúbal.

ALENTEJO

Sado Estuary

This large wetland is currently a nature reserve which extends over 23,000 ha. The northern shore suffers from heavy development and ugly pollution, but the rest of the area has remained relatively untransformed, apart of course from forestry and extensive agricultural activity, centred on rice production. The whole area is biologically rich and varied, forming one of the best sites for birdwatching in Portugal. There are breeding, passage and wintering waterbirds, as well as a remarkable variety of breeding and passage passerines.

Location *Geographical coordinates (Setúbal): 38°31′19.00″N, 8°53′06.00″W*

This site lies about 50km south-east of Lisbon on the west coast of

Portugal, close to the city of Setúbal. Full details of the area can be found on the following maps: 1:100,000: 39; 1:50,000: 39-A, 39-C; 1:25,000: 455, 465, 466, 467, 475 and 476.

Description

The landscape is varied. The estuary is 35km long, from the mouth, near Setúbal, to Alcácer do Sal, in the south-east. It is very wide, but very shallow and when the tide is low, most of its surface becomes exposed as intertidal mud.

The river banks are covered in places with large expanses of saltmarsh and reeds. There are extensive complexes of what were originally saltpans ('salinas') all around the estuary, particularly on the northern shore, to the west of Zambujal and on the southern shore, to the west of Alcácer. Most of them are no longer used for salt production: they are either abandoned or have been converted to fish ponds. Small freshwater pools, such as Açude da Murta, occur throughout the region and many of them are also surrounded by large patches of reeds and willows.

The wooded areas are very important, extending as they do for many kilometres. On the northern shore, they are mainly formed by dense Cork Oak forest, together with Maritime Pine, while on the southern shore Umbrella Pine is by far the dominant species, though Olive, eucalyptus and other trees occur.

The peninsula of Tróia forms the western limit of the estuary and separates it from the Atlantic. It consists almost entirely of sand and is densely vegetated with bushes, such as those of acacia and plantations of Maritime Pine and eucalyptus. The northern part of the peninsula has been built up with tourist resorts.

The sapal da Carrasqueira is a small treeless peninsula of land that extends into the estuary northwards of Comporta. It lies immediately to the east of the Tróia peninsula and is surrounded by a dyke in order to prevent flooding; rice is cultivated extensively here. At the northern extreme there are mainly pastureland areas and also some pools and reedbeds. Outside the dike there are vast saltmarsh extensions.

Birds

The Sado estuary has a varied avifauna at all times of the year, but especially so outside the breeding season, when the migrant and wintering waterbirds become numerous. The area is of international importance for its wintering populations of wildfowl and shorebirds. Noteworthy are the large numbers of Teal, Pintail, Wigeon, Mallard and Shoveler. The wintering population of Black-necked Grebes is the largest in Portugal, and the same applies to Red-breasted Merganser, of which up to 350 winter regularly in the estuary. The number of Cormorants that can be seen outside the breeding season is high, reflecting perhaps the increasing use of saltpans as fish ponds and the increase in their north-western European population.

Additionally, there are large numbers of wintering shorebirds, particularly Avocet, Grey Plover, Ringed Plover, Dunlin, Redshank, Curlew, Black-tailed and Bar-tailed Godwits and Snipe. Less numerous are Greenshank, Spotted Redshank and Turnstone. There is also a wintering population of Black-winged Stilts. Passage brings Ruff, Curlew Sandpiper, Knot and various terns.

Bluethroat

Other common wintering waterbirds include White Stork, Grey Heron, Little and Cattle Egrets, Glossy Ibis or Greater Flamingo (although many oversummer as well), Spoonbill, Black-headed and Lesser Black-backed Gulls and Sandwich Terns. Among the smaller passerines, a few interesting species occur in winter in small numbers: Water Pipit, Bluethroat, Penduline Tit, Spanish Sparrow and Reed Bunting.

During the breeding season, the estuary is somewhat quieter, due to the absence of large numbers of waterbirds. However, the variety of breeding birds is very interesting. White Stork, Purple Heron, and Cattle and Little Egrets are all fairly common. Black-winged Stilts are abundant and can be seen in most saltpans and pools. Other breeding waterbirds include Kentish Plover, Yellow-legged Gull and Little Tern.

The woodlands hold Tawny Owl, European and Red-necked Nightjars, Green, Great Spotted and Lesser Spotted Woodpeckers, Bee-eater and Hoopoe. These areas are also interesting for their wide variety of breeding passerines. Apart from the usual, more common birds, several noteworthy species are particularly abundant here, like Woodlark, Crested Tit, Nuthatch, Short-toed Treecreeper, Woodchat Shrike, Spotless Starling, Azure-winged Magpie, Serin, and Cirl Bunting. Less common and harder to see are Great Spotted Cuckoo, Western Bonelli's Warbler, Iberian Chiffchaff, Redstart, Golden Oriole, Rock Sparrow or Hawfinch.

The reedbeds scattered all over the estuary are good places to look for Water Rail, Cetti's, Reed and Great Reed Warblers. The dunes in the western part of the area hold Thekla Lark and Dartford Warbler.

Both Swift and Pallid Swift are common breeding species nearby, while Alpine Swift is a passage migrant.

Surprisingly, there are few birds of prey. Black-winged Kite, Buzzard, Marsh Harrier, Booted and Short-toed Eagles and Kestrel are seen regularly in the area throughout most of the year, but the other raptors, like Hen Harrier, Merlin, Peregrine Falcon and Osprey, are mainly non-breeding visitors. There are also regular sightings of Bonelli's Eagle.

As happened in some other coastal wetlands, a few exotic species have 'discovered' the rice fields and ditches around the estuary. Common Waxbill is numerous and Yellow-crowned Bishop can be regularly found in some places. There are also records of Red Avadavat and Tricoloured Munia although their breeding status is still uncertain.

Accommodation

Accommodation is not a problem. Only 40km from Lisbon with all its facilities, Setúbal has also a wide variety of hotels, suiting all pockets. It is also possible to stay in the Tróia complex but it should be kept in mind that prices can be high in summer. At Comporta there are houses to rent and for those with a smaller budget there are camp sites at Outão (near Setúbal), Gâmbia and Alcácer do Sal.

Strategy

Setúbal is the best starting point to visit this site and travel from Lisbon to Setúbal is straightforward using the A2 or A12. Due to its large size, the Sado estuary takes two or three days to explore thoroughly, but about 70 species can be seen in a visit of one day at all times of the year, with up to 100 being seen on good winter days. Note that there are many access restrictions to several areas around the estuary and only the sites where access is permitted will be mentioned.

a) Southern (western) bank

The southern bank of the Sado estuary can be explored from the N253-1 on the Tróia sandbar and on the N253 which runs from Comporta to Alcácer do Sal.

a1 – Tróia to Comporta

Tróia is a large tourist complex at the mouth of the Sado river. To get there, it is probably wisest (and often very rewarding) to take the car ferry in Setúbal. The crossing takes about 30 minutes and the ticket costs (2013 prices) €12.90 for a car including the driver, €3.20 for an adult passenger and €2.00 for children under 10 years. During the winter, the crossing may allow for sightings of Red-breasted Merganser, Common Scoter, Mediterranean Gull and Razorbill, along with Bottle-nosed Dolphins, which are still fairly common in this area. If you don't want to spend money on the boat or if there is a long queue to cross (it can happen in summer months or at weekends) there is an alternative solution – to drive around the estuary from Setúbal to Alcácer do Sal and then take the N253 there towards Comporta and Tróia. Note that the ferryboat no longer stops at the very tip of the sandbar; the dock is now located about 5km from the tip.

If you are staying at Tróia you can easily visit the so-called Caldeira de Tróia which is a small bay located on the east side of the Tróia peninsula. You can access it on foot or via the road leading to the Roman ruins. It usually has many waders, including Curlew, Sanderling and Turnstone. Occasionally, other interesting species put in appearance, like Spoonbill and Mediterranean Gull.

There is only one road leading south from Tróia (N253-1); it leads to Comporta (about 15km to the south) and from this road one can obtain good views over the estuary. Look for Greater Flamingos, waders and wildfowl, particularly Red-breasted Mergansers. At Comporta there are large expanses of saltmarsh and exposed mud, which often hold large numbers of waders. Avocets can often be seen here, White Storks nest in the village and Thekla Lark can be found along the road to Comporta beach and on the surrounding sand dunes.

Sapal da Carrasqueira lies north of Comporta. Access is possible from the Carrasqueira-Comporta road. At Comporta find Rua das Palmeiras and from there proceed towards Possanco. After about 2km, veer left, crossing a small residential area and then taking a rough track northwards through the rice fields. Small reedy pools at the northern part of this peninsula hold Water Rail, Marsh Harrier, Penduline Tit and reed warblers. From the vantage point of the dyke around these rice fields it is possible to scan the estuary proper. The rice fields themselves are used as feeding grounds for Glossy Ibis, Spoonbill and Water Pipit. This is also a good place for Short-eared Owl in winter.

a2 – Comporta to Alcácer do Sal

The road forks at Comporta, with one road (N261) leading away from the estuary (south to Lagoa de Santo André) and the other (N253) running westwards towards Alcácer do Sal. The latter road is the key one to use in order to explore this southern sector of the estuary. Between km 7 and 8 (closer to the latter) near the tiny village of Murta, there is a small dam rather well concealed; it is called Açude da Murta and it lies about 100m to the right of the road. This place used to hold a large colony of Cattle and Little Egrets. Although the colony has now disappeared, it is still a good place to stop (careful with the sand when parking). Marsh Harrier breeds and Booted and Short-toed Eagles are regular, mainly during the breeding season, while Great Spotted Cuckoos often display in late winter and early spring. In winter, this is a good place to see Penduline Tit and some waterfowl.

Carrying on to Alcácer do Sal, there are many areas of saltpans lying to the north of the road close to the estuary. These saltpans often hold interesting waders, especially at high tide. They can be reached by following one of the many tracks to the northern side of the road, but care should be taken on mud or deep sandy tracks. Closer to Alcácer, the former saltpans of Batalha are worth visiting although access is not easy and can even be tricky, especially after rain. Great White Egret is now regular in the nearby fields and Black-winged Kite also turns up. Sometimes Caspian Terns can be seen flying around in winter. To get there you must take a sandy track that leaves north from the N253 between km 20 and 21. This track passes nearby to a small rural airfield and proceeds to the right towards the river.

The area that extends to the south of this road forms the largest patch of Umbrella Pine in Portugal. The variety of birds is low, but this is a good spot for Azure-winged Magpie.

The town of Alcácer do Sal has large numbers of breeding White Storks, mainly in the church towers and around the castle (which has been converted into a luxury hotel).

b) Northern (eastern) bank

The northern (eastern) bank of the estuary can be explored from the

N10 starting at Setúbal up to Águas de Moura and from the N5 or IC1 after that village until Alcácer do Sal.

b1 – Setúbal to Águas de Moura

Setúbal has a large colony of Pallid Swifts, best looked for in the old part of town. If you are there you should not miss it. Departing from Setúbal eastwards on the N10, along the first 15km there are many complexes of old saltpans off to the south side of the road. Only the more accessible complexes will be mentioned here. Close to Setúbal, and surrounded by factories and heavy industry, the Mitrena saltpans can be interesting at high tide as they sometimes hold large numbers of waders. There are also mudflats around here. From the N10, at Alto da Guerra you must follow the signs to Mitrena. This area can be also be reached from down town Setúbal following the N10-4 road that leaves eastwards from the ferryboat terminal and runs between the railway line and the river – Mitrena is reached after 9km.

Not far from Mitrena there is the village of Praias do Sado. Close to its south-western tip there is a large wetland just by the railway line which usually holds a large number of aquatic birds. Care should be taken in parking and crossing the railway. Proceeding through the village, on reaching the eastern end, near a sharp curve, there is a rough track to the right which allows access to another saltpan complex. Note that some of these saltpans are fenced and closed for fish farming purposes. Do not enter without permission.

Close to the small village of Mourisca there are also some saltpans and marshy areas that can hold a few waders. There are some nature

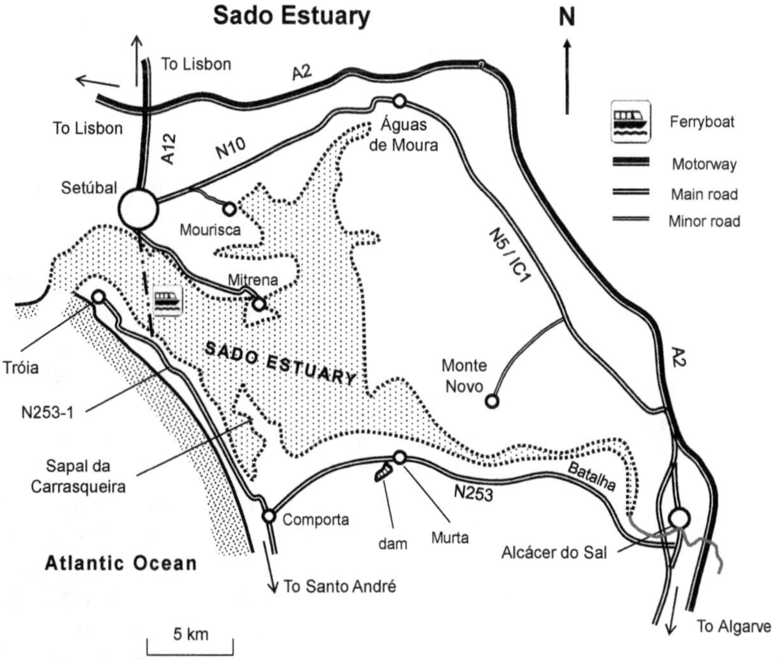

reserve facilities there at an old tidal mill, which are open from Wednesday to Sunday from 10.00 to 18.00. There is also a wooden birdwatching hide nearby. Mourisca can be reached from the N10 a few hundred metres east of Alto da Guerra, veering right just before km 45 following the signs to 'Mourisca'. After about 700m one must proceed left following the signs to 'Moinho de Maré da Mourisca'. About 500m after crossing the railway line, at a small roundabout, turn left.

Back at the N10, further east, after passing the railway station of Águas de Moura and close to km 54, there is another road to the right opposite a signpost reading 'reserva natural'; this road, called 'Estrada do Zambujal', passes under the railway line and leads to an area of reedbeds, rice fields, abandoned saltpans and saltmarsh. The road leads to a metal bridge over the river which as of 2013 has been closed for security reasons. However, good birding is possible even without crossing the bridge. Here there are Black-winged Stilt, Iberian Yellow Wagtail and Great Reed Warbler in summer. Outside the breeding season there are lots of interesting waders, as well as Bluethroat, Dartford Warbler, and Penduline Tit. This is a feeding and roosting place for big numbers of Glossy Ibises. Caspian Tern and Osprey are regularly seen in winter fishing in the river. The list of rarities seen here is long and includes species like Lesser Yellowlegs and Long-billed Dowitcher.

b2 – Águas de Moura to Alcácer do Sal

From Águas de Moura take the N5 (also called IC1) southwards to Alcácer do Sal. The road passes through an area of dense woods of Cork Oaks and pines, which hold large numbers of forest birds. Many of the unsurfaced tracks that lead westwards are now closed. However it is worth exploring the ones that remain open in order to get away from the main road. However, remember your route through the woods, as it is disconcertingly easy to become disorientated by the multitude of similar-looking trees and sandy tracks. Note that most of the land is fenced off here and probably it will be difficult to leave the tracks to walk through the woods.

About 15km south of Marateca, just past Palma, there is a surfaced road to the right signposted 'Monte Novo', which leads to a small village close to the river. This is a very interesting area with a rich bird life and it is strongly suggested that one walks the few kilometres circuit around the village. There are extensive rice fields, which are worth exploring. Closer to the river shore, there are old saltpans, which hold large flocks of passage and wintering waders. The surrounding pine and Cork Oak woods hold a large variety of woodland birds.

Returning to the N5, another good possibility to explore the wooded area is to take the track to the right close to km 49. There is an open gate with two pillars written 'Herdade da Boa Esperança' and 'Herdade de Santa Maria de Espim'. This track can be driven without major problems – it has no exit and gives access to a very small village after about 5km. Although you cannot walk through the woods, regular stops can produce interesting birds, like Western Bonelli's Warbler, Crested Tit, Azure-winged Magpie and Hawfinch.

Lagoa de Santo André

This is one of the largest coastal lagoons in western Portugal. Although a very important wetland at all times of year, it became protected only after 2000. Its coastal location and its vegetation render it particularly important during migration time, with large numbers of migrating waders passing through. An annual ringing campaign here in autumn has revealed that the area is also used by large numbers of migrant passerines. Additionally, this is an important place for wintering waterfowl.

Location

Geographical coordinates: 38°06′01.00″N, 8°47′20.00″W

Lagoa de Santo André lies in the dunes of the western Alentejo about 120km south of Lisbon. Full details of the area can be found on the following maps: 1:100,000: 42; 1:50,000: 42-A; 1:25,000: 505.

Description

The lagoon itself covers about 600ha in winter, but the management regime produces a series of seasonal changes in water level, salinity and productivity, which are reflected in a wide variety of bird life. There is a main water body which forms the lagoon proper; on the south-western side, concealed among the dunes are some smaller lagoons, known locally as 'poços'.

An important part of the eastern shore is covered by aquatic vegetation, formed by reedmace, reeds and willows. The lagoon is separated from the sea by a large chain of sand dunes covered mainly with Maritime Pines and low dense scrub. Although the sandbar is usually closed, it is seasonally opened by man, usually once a year. Further inland, the soil is still composed largely of sand and considerable woodlands exist, dominated mainly by plantations of Maritime Pine and Cork Oak, with few cultivated areas.

Human activities vary according to the time of the year and like many other wetlands in Europe, it is under threat from a variety of sources. Pressure resulting from tourism can be heavy in summer, although people concentrate largely along the northern part of the lagoon, which lies close to the beach and around the small village of Costa de Santo André. Fishing with nets is a very common and authorised practice.

Birds

The richest months are March, April, August and September, as numbers of species are augmented by passage birds, many of which are trans-Saharan migrants, though each season has its own specialities.

In winter, the lagoon holds a wide variety of dabbling and diving ducks, namely Mallard, Teal, Shoveler, Pintail, Gadwall, Pochard and Tufted Duck. One of the local specialities is Red-crested Pochard, of which up to five hundred can be seen at times. Coot counts can exceed 10,000. Among them is sometimes the odd Red-knobbed Coot. Water Pipits and Penduline Tits are also regular in winter. Glossy Ibis, Spoonbill and Purple Swamphen are frequently seen. Other interesting wintering species have included Black-necked Grebe, Squacco Heron, Great White Egret, Greater Flamingo, Short-eared Owl and Richard's Pipit.

Spring comes early here and by mid-February hirundines are common. Great Spotted Cuckoos are seen from mid-February onwards and warm anti-cyclonic conditions in March can bring early passage of Garganeys, Short-toed Larks, Yellow Wagtails, singing Sedge and Willow Warblers and Woodchat Shrikes. Spring arrivals are much dependent upon weather conditions and throughout April and May falls of migrants may occur; these have included White-winged Black Tern and Roller. By this time the reedbeds are vibrating with the songs of Savi's, Reed and Great Reed Warblers and at dusk Little Bitterns can be heard 'barking'. Purple Herons are present throughout the breeding season.

In summer locally breeding Black-winged Stilts, Bee-eaters and hirundines can be common. There are also a few pairs of Little Terns. Hobbies breed nearby and can often be seen scything through the masses of hirundines at dusk. Crested Tit, Hawfinch and Cirl Bunting are common in the surrounding woods. Other resident landbirds include Stone-curlew, Crested Lark, Azure-winged Magpie and Spotless Starling.

Post-breeding movements begin in July with a few early returning Holarctic waders and the abmigration of ducks and Coot from Spain to moult at the lagoon. From August to November, heavy passage has been recorded and most trans-Saharan migrants can be seen in variable numbers.

Rarities have included Sociable Plover, Pectoral Sandpiper, Lesser Yellowlegs and Citrine Wagtail.

Accommodation

There is a small hotel in Costa de Santo André and a camping park. There is also a hotel at Vila Nova de Santo André. At nearby Lagoa de Melides there is a vast choice of rooms and houses to rent. Besides, a variety of accommodation can be arranged in most of the villages and small cities within half an hour to one hour's driving range. It is also possible to reach the area in 2 hours from Lisbon or 1.5 hours from Setúbal and of course these cities offer a wide variety of pensions and hotels.

Strategy

From Lisbon or the Algarve, take the A2 motorway and exit at Grândola. From there follow the IC33 towards Sines. Leave at either the Cruz de João Mendes or Santiago do Cacém nodes and follow the signs to 'Lagoa de S.to André'. Upon reaching Brescos village, you are entering the site. At Brescos, the road forks: one road leads westwards to the beach and the other leads southwards to the far end of the lagoon.

a) Northern sector (west of Brescos)

The north shore of the lagoon can be easily explored from Costa de Santo André even on foot. Start at the terminal sector of the lagoon close to the beach. Although this is a very disturbed area, especially in summer, the small islands that exist there usually holds a few pairs of Little Terns and besides there are sometimes some Greater Flamingos around.

Getting back on the road to Brescos, after passing the turn to the campsite, veer right on a track signposted 'Monte do Paio' – at this place, which is reached after about 0.5km, there is a facility of the nature reserve run by Instituto da Conservação da Natureza e das Florestas (ICNF – the government body concerned with nature conservation) and it is possible to get a map of the area (the opening hours are: 10.00 to 12.00 and 14.00 to 16.30).

Monte do Paio is the starting point of an excellent marked foot path that runs southwards along the lagoon shore until one reaches Casa do Peixe ('fish house'), which is in fact a set of two ruined houses about 2km from the beginning of the pathway. This trail crosses some of the typical habitats of the area and allows observation of a wide range of species. From the geodetic mark that lies half way there are superb views over the lagoon with thousands of coots and ducks in the winter. Savi's Warbler can be heard singing from the rushes and Iberian Yellow Wagtail is common. From Casa do Peixe it is also possible to obtain good views over the lagoon and to explore some areas of aquatic vegetation, although most of the reedbeds are strictly protected and access is forbidden. Savi's and Reed Warblers breed nearby. It should be noted that this trail might not be passable during winter months, depending on water levels.

b) Eastern and southern sectors (south of Brescos)

By leaving Brescos southwards it is possible to explore the eastern shore of the lagoon. At the village of Brescos on the M544 veer southward following the signs to 'V. Nv. S.to. André' and 'Sines'.

Since the creation of the nature reserve, a few tracks have been closed to vehicles. However, they can be traversed on foot. That is the best option in order to explore the flooded areas of the eastern sectors. These can be approached from paths that start on the road from Brescos to Vila Nova de Santo André. One of them starts at a gate which marks the beginning of the 'Moinho Novo' path; it brings one through pines and Cork Oak to the south-eastern corner of the lagoon. Walk northwards to a small cement bridge. The area is very good in summer for breeding warblers and in winter for Penduline Tits and wildfowl. By following the track north, it will bring you to the heart of the main reedbeds; depending on water level and season, these can be excellent for waterfowl, Spoonbills, Glossy Ibises, Purple Swamphen waders, terns and passerines.

The southern and south-eastern side of the lagoon can be explored from the road that leads to Porto das Carretas (Vacaria) beach. Coming from Brescos, you should veer right after about 5km towards Monte Velho and beach (look for signs to 'Praia do Porto das Carretas'). From this road it is possible to have good views over the flooded fields and the pastures of the south-eastern tip of the lagoon – the water edge can be rather close to the road or very far away, depending on precipitation levels. These areas are used by good numbers of herons and storks including Great White Egret. About 900m after taking this road, there is a sewage works pool (ETAR) on the left. This is worth checking, as there are often a few ducks (the place is a regular haunt of Red-crested Pochard), grebes and coots.

Before reaching the beach, about 2.5km from the turning and just before the road dips down towards the sea, there is a sandy track to the right that goes north over/along the dunes. Driving is forbidden but it is possible to walk; this very sandy track leads to the small lagoons ('poços'). In this case it will lead to Poço do Parral and Poço dos Caniços which appear on the right side after about 10 min walk – a wooden structure has been built here and can be used as an observation point. The number of birds here can be highly variable, but usually these are very good for waterfowl including Red-crested Pochard. Sometimes there are Ferruginous Ducks and Scaup. The calling of Stone-curlews is constant in the surrounding area and Hobby is regularly seen hunting here.

Serra de São Mamede

Close to the city of Portalegre, serra de São Mamede forms the highest point south of the river Tagus. This area marks the transition between the plains of the south and the mountains of the north. As a result, the landscape is varied and bird life is very rich, as both northern and southern species can be found here.

Location

Geographical coordinates (Castelo de Vide): 39°24'54.00"N, 7°27'15.00"W

This site lies in the northern part of the Alentejo province, about 200km north-east of Lisbon. Full details can be found on the following maps: 1:100,000: 28 and 29; 1:50,000: 28-B, 28-D and 29-C; 1:25,000: 325, 335, 336, 347 and 348.

Description

A hilly area that stands out from the surrounding Alentejo plains, serra de São Mamede rises to 1025m above sea level. Due to altitude, the climate is clearly cooler and wetter than in other parts of the province and this has a marked effect on flora and fauna. The whole place has been declared a natural park.

Because of the juxtaposition of hills and plains, and because of its location in the transition between northern and southern Portugal, this is a region of contrasts.

To the north, around Castelo de Vide and Marvão, there are extensive Cork, Holm and Pyrenean Oak woods (the latter species reaches its southern limit here), as well as scrubland, consisting mainly of *Genista* and Sun Rose. Granitic rocky outcrops dot the landscape. There is a dam (barragem da Póvoa) in the north-western tip of the area, near the village of Póvoa e Meadas.

Further to the south, lies the serra de São Mamede proper, covered largely by Maritime Pine and eucalyptus plantations, but there are also several Sweet Chestnut groves and a few oak woods.

Finally, the southern part consists of what might be considered the typical Alentejo landscape with cereal fields and the montados (open woodland formed mainly by Cork and Holm Oak).

Birds

This is one of the richest areas in Portugal in terms of landbirds: over one hundred species breed and many others occur as non-breeding visitors.

The variety of passerines is especially striking with most Portuguese species present here, namely six species of lark, including Calandra, Short-toed and Thekla, five species of hirundines, four species of sparrows and an array of warblers including Spectacled, Subalpine, Western Orphean, Melodious, Dartford, Western Bonelli's and Cetti's, as well as Fan-tailed Warbler. Blue Rock Thrush is fairly common throughout this region and can be found both in villages and in rocky outcrops.

Other interesting passerines include Black Redstart, Southern Grey and Woodchat Shrikes, Firecrest, Crested Tit, Short-toed Treecreeper, Golden Oriole, Spotless Starling, Azure-winged Magpie, Serin, Hawfinch and Rock and Cirl Buntings. Dipper has bred. The rare Black Wheatear used to breed until at least the year 2000 and although there are no recent sightings it might still occur. In winter there are Alpine Accentors.

Birds of prey are quite varied, with three species of vulture, three species of harrier, and all large eagles, along with three kites, including Black-winged, buzzards and Sparrowhawk. Apart from diurnal raptors, other breeding species include White and Black Storks, Quail, Little Bustard, Stone-curlew, Eagle Owl, Great Spotted Cuckoo, Red-necked Nightjar, Swift, Pallid and maybe Alpine Swifts, Bee-eater, Hoopoe and Iberian Green Woodpecker (race *sharpei*). Rollers used to occur in the area but they have become rare. Waterbirds are less numerous but a few interesting species can be found namely Little Egret, Little Ringed Plover, Kingfisher and Water Pipit.

Accommodation

There are two hotels at Castelo de Vide, one at Portalegre and another one at Marvão. Furthermore, there are pensions in all these towns. Finding a place to stay should not be a problem, except possibly during summer weekends, New Year and on the second weekend of November, when there is the chestnut fair at Marvão.

Strategy

This area is particularly interesting during the breeding season (March to June), but other seasons can also prove rewarding.

From Lisbon, take the A1 northwards, leave where signs indicate

'Torres Novas' and then proceed eastwards on the A23 through Abrantes leaving at Mouriscas, after that on the N118 through Gavião and Alpalhão and finally on the N246 until Castelo de Vide. Another option is to leave Lisbon eastwards on the A6 until Estremoz and then take the IP2 northwards until Portalegre and at last the N246 to Castelo de Vide.

a) Castelo de Vide area

This picturesque town can be used as a base for exploration is worth a visit. In particular, visit the castle and climb the tower (free of charge), as the views are far-reaching and Blue Rock Thrush is usually around. Crag Martins sometimes breed here.

The viewing point of Senhora da Penha (on the hill facing Castelo de Vide from the south) provides superb views as well as a few interesting birds, including Rock Bunting, while the nearby pinewoods have Crested Tit and Firecrest.

An interesting area lies to the north of the town, towards Póvoa e Meadas. Leave Castelo de Vide westwards towards Lisbon and after 4km turn right following the signs to 'barragem'. The road is narrow but traffic is light so it is worth stopping wherever possible to look for birds, but be careful as always. Interesting species along this road include Great Spotted Cuckoo, Thekla Lark and various *Sylvia* warblers. After a few kilometres, the reservoir of the barragem da Póvoa appears on the right side. It is possible to park on the road side and scan for waterbirds. There are also a few tracks that make it possible to approach the water, although waterbirds are usually scarce here. However, it is also worth

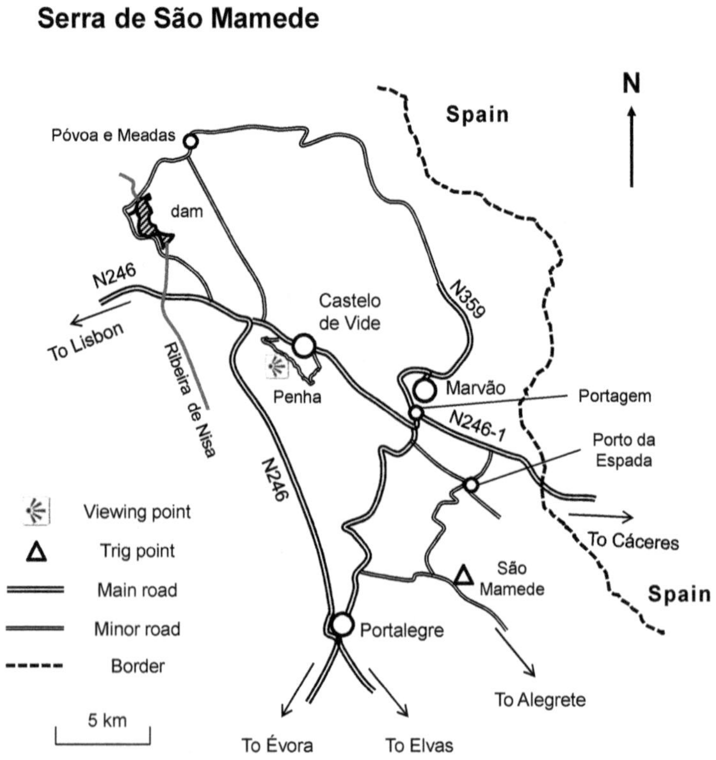

scanning for raptors: Black Kite is frequent in summer and Short-toed Eagle is regular in the surroundings.

Finally, the dam wall is reached and there is a car park just after crossing the wall. The large eucalyptus trees here support several nests of White Stork, with Spanish Sparrows breeding under them. The nearby areas hold several woodland birds, while the dam wall itself is often used by feeding Crag Martins and Red-rumped Swallows. Proceeding through Póvoa e Meadas and taking the road that leaves eastwards to Marvão, one enters a region covered with Holm Oaks where Azure-winged Magpies are numerous. There are a few small bridges along the road which are all worth inspecting for Red-rumped Swallow. After a few kilometres, the landscape opens up and you are entering Calandra Lark territory, which can easily be heard singing in spring. Other open-country species can be found here, including Little Bustard and Thekla Lark. The views reach far into Spain and a stop here might produce large soaring birds such as vultures and large eagles.

b) Marvão area

To the east of Castelo de Vide lies the historical village of Marvão, another interesting spot. The castle has breath-taking views and is an excellent place to see Alpine Accentor in winter, as well as Crag Martin, Blue Rock Thrush and Rock Bunting year round. Until recently Black Wheatear was regular here. To get there, leave Castelo the Vide eastwards on the N246-1 following the signs to 'Espanha' (Spain). After about 7km, turn left and proceed on the N359, The huge rock of Marvão is now clearly visible ahead. Roadsigns reading 'Marvão' show the way to this village.

Driving back down, the village of Portagem (lying just 6km drive from Marvão, at the base of the hill), is worth a stop. There is a nice park along the river which is also worth a walk to look for passerines. Short-toed Treecreeper is usually guaranteed here.

From Portagem, the N246-1 leaves eastwards and leads to the Spanish border, which lies 8km away; just before reaching the border, it passes beneath a huge rock which holds a few breeding pairs of Griffon Vultures (easy to watch from the road). Other birds that can be seen here include Red-rumped Swallow and sometimes Alpine Swift.

At Porto da Espada (which lies midway between Portagem and the border) there are Sweet Chestnut groves, which are good for several passerines, including Rock Sparrow. From here a minor road leads to the top of serra de São Mamede – these hills are mainly covered with Maritime Pines and birds are less varied, but in the top areas it is possible to find Crested Tit and Rock Bunting.

Elvas Plains and Upper Guadiana River

The Elvas area was formerly regarded as one of the best places in Portugal for open country species. Nowadays, the situation has changed markedly; intensification of agriculture has led to increased disturbance and habitat changes, furthermore, a motorway has been built across the area. Nevertheless, this region holds one of the richest bird communities in the country and it is still possible to find here most of those species one would expect to find in southern Portugal, including soaring birds, steppe birds, waterbirds and small passerines.

Location *Geographical coordinates (Elvas): 38°53′41.00″N, 7°09′60.00″W*

The site lies due east of Lisbon, in the eastern Alentejo, close to the Spanish border, where the river Guadiana enters Portugal. Full details of the area can be found on the following maps: 1:100,000: 33 and 37; 1:50,000: 33-C, 33-D and 37-A; 1:25,000: 400, 401, 413, 414 and 428.

Description Elvas is a medium-sized town that has grown markedly in the last few years. The old part of the town is still surrounded by a large wall which was built to protect the inhabitants from attacks in former times.

Not far from Elvas, the frontier is marked by the Guadiana and by its tributary the Caia. Both rivers have flat and very fertile valleys. The riverine vegetation is very rich, with many Poplars and eucalyptus, dense bushes and some reedbeds. The larger tributaries have similar vegetation, while the small streams are lined with *Genista* and Oleander, especially around Campo Maior, which lies a few kilometres to the north-east.

Agriculture, which is both intensive and extensive, predominates and there are large irrigation systems, with huge expanses of sunflower and corn fields, as well as Olive groves (both industrial and traditional). Away from the river, wheat is the dominant cultivated crop, but the existing woodland groves, consisting mainly of Olive and Holm Oak, are an important feature of the landscape. There are also large areas of pastures which are used for cattle grazing, especially west of Elvas.

The barragem do Caia, built directly on the river Caia a few kilometres to the north of Elvas, forms the largest reservoir in this region, while small dams have been built here and there, mainly for water storage and irrigation.

Birds This area is particularly important for breeding birds, being most interesting in spring and early summer. Widespread and easy to find are White Stork, Bee-eater, Hoopoe, Southern Grey and Woodchat Shrikes, Spotless Starling and Serin. Sadly steppe birds have declined dramatically in recent years, due mainly to intensification of agriculture, which has destroyed much of their natural habitat. Little Bustard and Stone-curlew can still be seen, but Great Bustard has become very scarce and Black-bellied Sandgrouse is apparently extirpated. The Elvas plains also used to be a good area for Roller, but this species seems to have also declined in recent years.

Breeding birds of prey are not numerous but there is an interesting assortment of them: Montagu's Harrier, Buzzard and Kestrel are the only common species, while Black and Black-winged Kites, Short-toed and Booted Eagles occur in small numbers. Lesser Kestrel does not appear to breed here anymore, although it is still widespread in the Spanish city of Badajoz, just across the border and a few are seen on occasion hunting on the Portuguese side.

The variety of passerines is large. There are five species of lark breeding: Calandra, Short-toed, Wood, Crested and Thekla. The latter two are common, giving an excellent opportunity to compare both species. Crag Martins and Red-rumped Swallows have a very localised distribution but can often be found near bridges. The list of breeding warblers is also interesting: Fan-tailed, Cetti's, Great Reed, Melodious and Sardinian Warblers are all fairly common and easy to find. Other breeding passerines include Black-eared Wheatear, Short-toed Treecreeper, Golden Oriole, Azure-winged Magpie and Rock Sparrow.

The Elvas region is also noted for the presence of exotic birds. Common Waxbill is widespread in reedbeds and riverine vegetation along the Guadiana and Caia valleys. The Red Avadavat apparently has colonised the region from Spain and is easier to find here than anywhere else in the country; it occurs in the same habitats as the Common Waxbill, although the two species do not normally form mixed flocks. Yellow-crowned Bishop has also been recorded in the area.

Other landbirds which can be seen during the breeding season include Quail, Great Spotted Cuckoo, Bee-eater and Penduline Tit – breeding of the latter was proved in 1994 but might happen only irregularly. Waterbirds can be found mainly at the dams (of which Caia is the largest). There are breeding Gull-billed and Little Terns, Collared Pratincole, Little Ringed Plover and Black-winged Stilt. Purple Swamphen has been recorded and might breed. Egrets are plentiful, the most numerous being Cattle and Little, but several other species have been recorded, either during breeding season (Little Bittern and Purple Heron) or as non-breeding visitors (Great White Egret). In mid-summer, the area has large post-breeding concentrations of White Storks (up to 1000 birds). Black Stork occurs on passage. The riverine vegetation is often used by migrant passerines, while the Caia reservoir attracts waders on passage.

In winter, many breeding birds are absent. Nevertheless, it is possible to find interesting residents and non-breeding visitors. Large flocks of Little Bustards often occur and Black-winged Kite is fairly common. There are also four species of larks still present. Winter visitors include Crane, Red Kite, Osprey, Hen Harrier and waterfowl.

Accommodation

There are several hotels in Elvas and at least one in Campo Maior. Furthermore, there are several cottages which offer accommodation outside the town.

As the area lies on one of the main roads between Spain and Portugal, there are a lot of restaurants both in Elvas and along the main N4 road.

Strategy

Winter and spring are probably the best seasons to look for birds in this region.

Elvas town is the starting point from which to visit the area. The surrounding countryside is worth exploring in all directions. The best sites are all within 20km of Elvas and are easy to visit.

Lying directly on the Lisbon-Madrid road, it can easily be reached from Lisbon (via the A2 and A6) – the journey can be made in about 2 hours. This place is also strategically located for those entering or leaving the country by car. For those coming from Spain, the city is reached shortly after crossing the border at Badajoz.

a) South and east to the Guadiana valley

Leave Elvas on the N4 towards Spain. Immediately after passing an old castle on the right hand side, veer right on a minor road signposted 'Torre de Bolsa'. After passing some Olive groves, one enters an area of open fields, which have Montagu's Harrier, Little Bustard, and Roller, among other birds. Stone-curlews are best looked for in the evening or at dusk, when they are calling. The elusive Great Bustard has been seen here. After about 6.5km, the road ends at a T-junction, beside two houses. A rough track leads straight ahead through open fields, which have Crested and Short-toed Larks, to the river Guadiana, which here forms the border with Spain. The riverine vegetation is rich. Penduline Tits have bred here.

Returning to the T-junction, turn right (east) and proceed until a small dam with extensive reed mace appears on the right side. This is called Alfarófia (another way to reach this place is to leave the A6 at the Caia junction and take the minor road southwards for 1km). Although the place is fenced, it is possible to watch from the road or from the gate. Species here include Purple Swamphen (occasional), Cetti's and Great Reed Warblers and the exotic Red Avadavat and Yellow-crowned Bishop.

To the south of Elvas lies another interesting area called Ajuda. Look for the road to Olivença, which leaves from the N4 near the pousada and reaches the river Guadiana after about 10km. Just before crossing the river into Spain, look for the signs to 'Ponte da Ajuda' (an old broken stone bridge). The landscape is attractive here and the river is

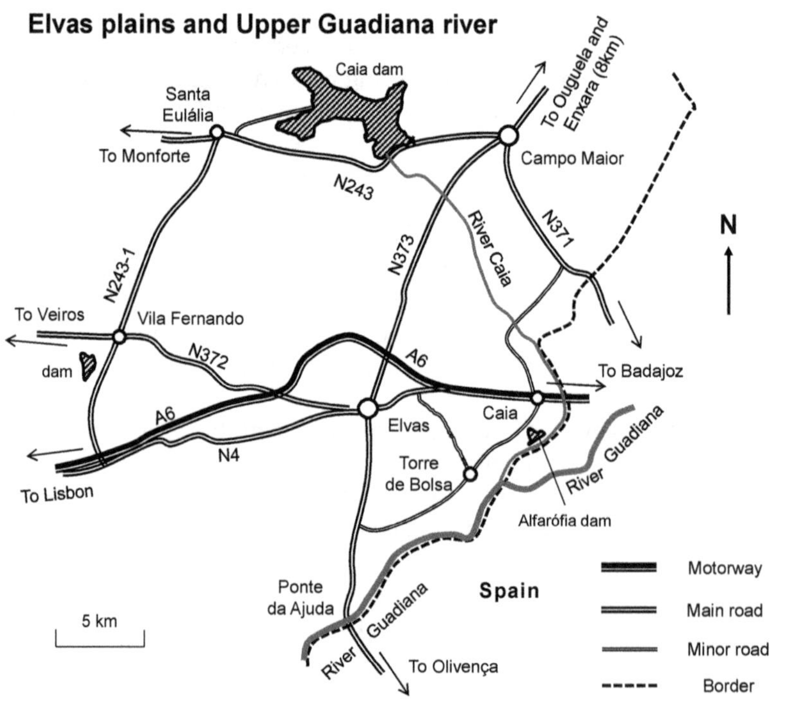

good for Little Egret and sometimes for migrating waterbirds, while the surrounding Holm Oaks just to the north are worth exploring for resident woodland passerines.

b) West to the plains around Vila Fernando

About 15km west of Elvas lies an area of open fields which is very good for steppe birds. Take the N372, which leaves Elvas directly from the aqueduct and proceed to Vila Fernando. From there, explore westwards (N372) and southwards (N243-1) taking one of the small tracks that leads into the open fields. This area is worth scanning for Great and Little Bustards and Stone-curlew. Other breeding birds here include Black Kite, Montagu's Harrier, Short-toed Eagle, Collared Pratincole (scarce), Roller and various passerines.

From Vila Fernando, it is also worth following the N243-1 southwards. About 2km along this road there is a reservoir lying directly on the right hand side of the road, which deserves an inspection.

c) North to Campo Maior and the Caia reservoir

The town of Campo Maior lies about 15km north-east of Elvas. There are two ways to get there: the quickest is along the N373 which leaves Elvas northwards and runs straight to Campo Maior (watch for Roller and Montagu's Harrier on the way). A longer but perhaps more interesting approach is to leave Elvas eastwards to the Caia border and from here take the minor road to the north signposted 'Campo Maior'. This route is quieter, hence better and safer for birdwatching. After about 2km from Caia, a small bridge appears. A stop is worthwhile as there are many reedbeds which hold Great Reed Warbler in summer. Further north, after passing a T-junction, veer right and proceed until you reach the next bridge, again with good patches of reedbeds. After a few more km, Campo Maior is reached.

From this town, it is worth proceeding northwards or westwards. To the north-east of Campo Maior, the N373 leads to Ouguela. Just before reaching Ouguela, veer right to Enxara and cross the narrow bridge of river Xévora. A small chapel appears on the right. This is a quiet place where parking is possible. It is worth exploring the riverine area and the fields and oak groves around the chapel. The castle of Ouguela (about 2km to the west) can be used in winter as a watching point in the late afternoon, as Cranes move to their roosts.

Back to Campo Maior, look for the N243 which leads westwards to the barragem do Caia. This road crosses open fields which are a regular haunt of Little Bustard, including large wintering flocks. Finally the barragem do Caia is reached. This is the largest reservoir in the region. A stop is possible just before crossing the dam wall. Gull-billed Tern breeds on the islands and can be seen during the breeding season while Little Bittern has been seen on the small pool just under the dam. However, the dam wall is not the best point to look for waterbirds. It is worth trying to approach the reservoir from the western side and this can be done by proceeding westwards on the N243, until the village of Santa Eulália appears. Just before entering the village, there is a small roundabout, close to the graveyard. A sign reading 'barragem do Caia' suggests a minor road to the right. This road (which is in bad condition, hence might not always be passable) ends after 4km very close to the western shore of the reservoir. Depending on season and water level, this place can prove disappointing or very interesting, but it is usually possible to find waterfowl, waders and other waterbirds.

Mora and Montargil

Geographical coordinates (barragem de Montargil): 39°03'18.60"N, 8°10'41.40"W

Lying in the transition zone between the provinces of Ribatejo and Alto Alentejo, Mora and its surroundings hold one of the best patches of woodland in southern Portugal. It consists mainly of Cork Oak groves that extend over many kilometres, but there are also some pine plantations. A few small rivers flow through the area and along these there is some riverine woodland, formed mainly by Willow trees. Slightly to the north, the barragem de Montargil forms the largest water body in the region.

This area holds an interesting variety of landbirds. The selection of raptors, both diurnal and nocturnal, is noteworthy; the commonest species are Black and Black-winged Kites, Buzzard and Booted Eagle. There are also a few pairs of Honey-buzzards, Sparrowhawks, Short-toed Eagles and Tawny and Eagle Owls. The woodlands hold many species of smaller landbirds, including Bee-eater, Wryneck, Woodlark, Western Bonelli's Warbler, Crested Tit, Woodchat Shrike, Hawfinch and Cirl Bunting. In the more open places there are a few agricultural fields, which seasonally hold feeding Cattle Egrets along with Crested and Short-toed Larks, Tawny Pipit, Black-eared Wheatear and Fan-tailed Warbler. The rivers and streams are good places to find Little Ringed Plover, Nightingale, Cetti's and Melodious Warblers and Iberian Chiffchaff.

This site lies about 110km east of Lisbon and the best route is via the Vasco da Gama bridge and then roads IC3, N119 and N251 past Coruche to Mora. It can also be reached from Évora in about 30 minutes. Leave Évora northwards on the N370, through Arraiolos and turn left at Pavia.

The best patches of woodland to look for landbirds lie around Cabeção and ribeira do Divor. For the former, leave Mora northwards on the N2 and turn right after crossing the bridge over ribeira da Raia, where a signpost reads 'Cabeção'. On reaching the village, it is possible to take the tarmac road that runs northwards towards Galveias through Cork Oak groves. Any of the unsurfaced tracks starting at this road can be used to explore the area. Note however that some might be fenced off and that there might be access restrictions in some places. For the latter, take the N2 southwards from Mora and veer right after 2km on a road signposted 'Ribeira do Divor' which will take you to a quiet area of woodland and gets close to ribeira do Divor.

Barragem de Montargil is worth inspecting. Its shores are muddy and shallow and attract waterbirds of all kinds. The N2 runs northwards along the western shore and provides good views over the water body. During the breeding season this place usually holds Great Crested Grebes and Little Ringed Plover, while Black Kite can be seen flying over the water. Little Egret does not breed here but it is often present, feeding along the shores. Migration and winter can bring waterfowl, Great White Egret, waders and terns. During the post-breeding period there are sometimes large concentrations of birds including White Storks, Spoonbills or Grey Herons among others. A stop at the dam wall usually provides Crag Martin. The surrounding areas hold Red-necked Nightjar.

Évora

Geographical coordinates: 38°33′58.00″N, 7°54′17.00″W

The historic city of Évora lies in the Upper Alentejo province and is worth a visit, both for its monuments and for its birds. Until recently, the surrounding plains were an ocean of wheatland and fallow fields, with a few oak woods here and there. However in recent years extensive Olive groves have been planted; these groves consist largely of new crops, industrially-farmed, and have led to marked changes in the landscape.

Despite of that a range of interesting steppe and open country species can still be found, including Little and Great Bustards (the last-named is very scarce), Montagu's Harrier, Lesser Kestrel, Stone-curlew and Calandra Lark. The rare Black-bellied Sandgrouse also occurs around here, albeit in very small numbers. In winter a few Cranes can be seen, along with Red Kites. The Olive groves hold good numbers of Crested and Short-toed Larks, Black-eared Wheatears and Tawny Pipits. The list of species commonly seen in this area includes summer visitors like Black Kite, Short-toed and Booted Eagles, Collared Pratincole, Gull-billed Tern, Great Spotted Cuckoo, Bee-eater, Red-rumped Swallow, Melodious Warbler, Woodchat Shrike and Golden Oriole and residents like Cattle Egret, Black-winged Kite, Short-toed Treecreeper, Southern Grey Shrike, Azure-winged Magpie, Spanish and Rock Sparrows and Serin.

Évora lies about 140km east of Lisbon and can easily be reached using the A6 motorway which crosses the Alentejo, linking the coast to the Spanish border.

The most interesting plains lie to the east and south of the city. The village of São Manços (20km south-east of Évora, on the road to Beja) is a good place to start exploring the area. The main roads are quite busy so the best option is to take one of the existing secondary roads, which provides access to most of the interesting habitats.

The N380 road that leaves Évora south-westwards towards Alcáçovas is also a good option to explore. Lesser Kestrels breed in old buildings not far from it and there is a rubbish dump ('aterro sanitário') just outside Évora that usually attracts good numbers of White Storks and Black Kites. The fields around the village of São Brás do Regedouro (20 km from Évora) can be interesting for open country species. It should however be noted that there are many fenced areas here and that permission should be asked before entering private land.

Barragem do Divor, which lies 15km to the north is reached using municipal roads (look for a sign reading 'Divor' near Évora's aqueduct). Interesting birds in this area include Little Bustard, Great Spotted Cuckoo and Rock Sparrow.

Mourão and Alqueva

Twenty years ago, Mourão was almost like a 'secret corner' in the eastern Alentejo, where relict populations of several endangered species still survived. The building up of the barragem de Alqueva, which was completed in 2002, brought great changes to this area, with one third of the Mourão county being submerged under reservoir water. Because of that, in just a few years, the birdlife of this region has changed dramatically: landbirds decreased and waterbirds increased. The juxtaposition of two very distinct habitats provides good opportunities for the visiting birdwatcher.

Location

Geographical coordinates (Mourão): 38°23'03.00"N, 7°20'12.50"W

The site lies about 50km to the south-east of Évora in south-eastern Portugal, close to the Spanish border. Full details of the area can be found on the following maps: 1:100,000: 41; 1:50,000: 41-A and 41-C; 1:25,000: 474, 483, 483-A, 492 and 493.

Description

Mourão is the name of a small town on the eastern bank of the river Guadiana, one of the three largest catchments in the country.

As a result of the construction and filling up of the huge barragem de Alqueva, there was significant habitat loss, along the river valleys, but on the other hand the reservoir created numerous islets which were soon colonised by waterbirds. The lake is so vast that it is impossible to see the whole extensiveness of it from one spot. The dam wall lies further south, closer to the town of Moura.

Away from the reservoir, the area is rather flat and open. Due to the poverty of the soil, the pattern of cultivated fields varies between years and large tracts are left undisturbed. The main crop is wheat. Many areas are used for cattle grazing and are therefore fenced off. Woodland consists mainly of Olive and Holm Oak groves. Human settlements are concentrated around villages, a few isolated houses can be seen here and there but many have been abandoned. South-west of Mourão, the village of Luz was newly built, in order to replace the old one which was 'swallowed' by the reservoir. Many Olive groves have been planted in recent years throughout the region.

Birds

Mourão has long been famous in Portugal for its steppe and open-country species. Although many of the birds that can be seen in this area occur over most of the Alentejo province, some of them are more easily found here than anywhere else in the country. One of the highlights of Mourão is Black-bellied Sandgrouse. This bird is still regular here and it is usually possible to see a few individuals. Stone-curlew is widespread and Great and Little Bustards occur, while in spring Calandra Larks can be found singing everywhere. Mourão is also quite good for soaring birds. Regular species during the breeding season include Black-winged Kite, Buzzard, Black Kite, Montagu's Harrier, Short-toed and Booted Eagles. Black Stork and Egyptian Vulture are scarce, while Cinereous Vulture is seen occasionally.

Summer visitors to this region include Great Spotted Cuckoo (best looked for in late February and March), Quail, Bee-eater, Roller, Short-toed Lark, Red-rumped Swallow, Rufous Bush Chat, Black-eared Wheatear, Subalpine and Melodious Warblers, Golden Oriole and Woodchat Shrike. Most nests of White Stork hold small colonies of Spanish Sparrows, which also breed in large numbers in wooded areas.

Black-bellied Sandgrouse

Winter can be exciting at Mourão, with several thousand Cranes roosting in the region. Flocks can often be seen feeding in the fields and oak groves or flying over in the evening, forming large 'letters' in the sky. Red Kites and Hen Harriers are also regular as non-breeding visitors, while Dotterel has been recorded on passage. Of course the resident passerines, such as Crag Martin, Southern Grey Shrike, Azure-winged Magpie, Spotless Starling and Rock Sparrow, also remain in the area.

The Alqueva reservoir has proved attractive for breeding waterbirds, namely Collared Pratincole, Gull-billed Tern and possibly Red-crested Pochard. Other species which can be seen include Little and Great White Egrets and a few migrating shorebirds. In 2011 and 2012, several young Ospreys 'imported' from Finland were released here, in an attempt to establish a breeding population and put the species back into the list of Portuguese breeding birds (the species formerly bred along the rocky coast in south-western Portugal).

Apart from the large Alqueva reservoir, there are a few small dams, which hold several breeding waterbirds, of which Little Egret, Little Ringed Plover and Black-winged Stilt are the most interesting ones.

Accommodation

If you plan to stay in this area, there are hotels in Monsaraz and Reguengos de Monsaraz and pensions in Mourão. There are also a few rural houses that have been converted for tourism purposes. If there are no rooms available, try Évora, which lies just 55km to the north-west and has a variety of accommodation facilities.

If you visit the Elvas area, then it is possible to reach Mourão in about one hour, returning to Elvas in the evening.

Strategy

This area is interesting throughout the year, except perhaps during summer months, when birds tend to be quiet and temperatures are unpleasantly high.

Starting from Évora, take the N18 and then the N256 eastwards through Reguengos de Monsaraz and proceed to Mourão village. From here, there are roads to many sites, allowing exploration of most of the area.

a) São Leonardo

Locate the graveyard at the eastern end of Mourão and take the road to the left, which is signposted 'Espanha' (Spain). This route (N256-1) runs eastwards to a hamlet called São Leonardo. The passport control here was withdrawn a few years ago and there are normally no problems crossing and recrossing the frontier. It is possible to park the car in São Leonardo and begin birdwatching in Portugal. However, it is probably better to cross the border and take the first surfaced road to the left, which leads through open fields to a small reservoir, where there is often a variety of waterfowl.

This area deserves a long careful circuit on foot, especially across the uncultivated fields, which form the main habitat for Black-bellied Sandgrouse and other steppe birds. Park on the other side of the reservoir and walk over the plains. The reservoir itself sometimes has a few waterbirds and in winter there are normally flocks of Cranes feeding and flying around, which should not be disturbed.

Returning to the N256-1, drive back and take one of the rough tracks to the left (southwards). These lead through open fields and Holm Oak groves and in summer can provide a good selection of birds. A few small reservoirs exist here and are worth inspecting. The Holm Oak

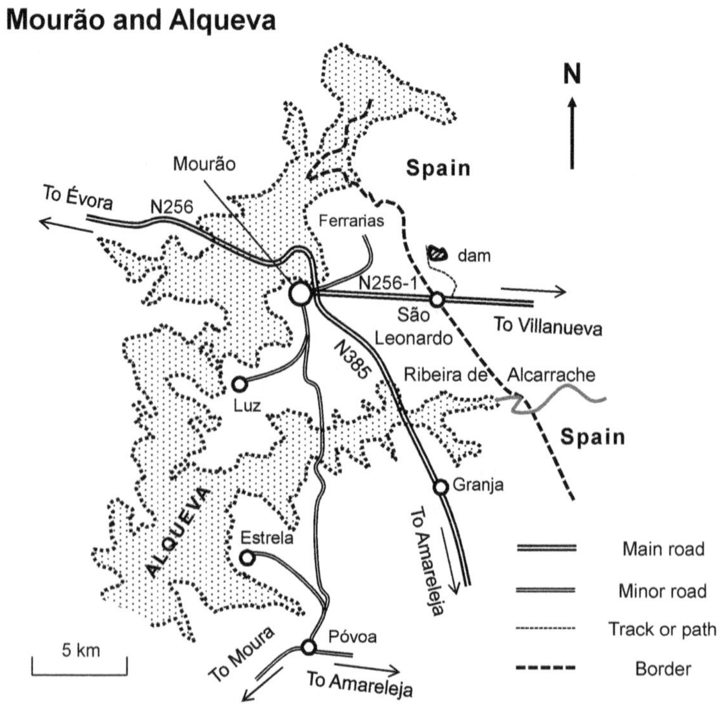

groves hold large colonies of Spanish Sparrows, although these tend to move from one year to the next and a careful search may be needed.

b) Ferrarias

This area lies to the north of Mourão and can be reached by following the signs to 'Campo de Futebol'. Past the football pitch and after about 5km the tarmac road ends and then you must veer left onto a dust track. This track crosses some open fields which are excellent for soaring birds of prey. Black-bellied Sandgrouse also occurs, but patient listening is essential. Little Bustard can also be seen here. The surrounding Holm Oaks hold Black-winged Kite, Great Spotted Cuckoo and several passerines.

c) Alqueva

The Alqueva reservoir extends to the north, west and south of Mourão, meaning that this town is surrounded by water from three sides. Several roads were 'cut' by the filling of the dam and can now be used to approach the water.

The best places to get close to the eastern shore are: Luz village, Estrela village and Alcarrache bridge.

Luz village lies to the south-west of Mourão. It is actually a new village that was built to replace the old Luz, which was submerged when the reservoir was filled. It is worth scanning the reservoir for ducks, egrets, coots or waders.

From Luz, take the narrow road eastwards and then turn right towards Póvoa; after 2km the bridge over ribeira de Alcarrache allows a good view over the eastern part of the lake – however, it is forbidden to stop on the bridge itself, so it is better to stop at one end and walk. This is a good spot to look for Cranes in winter. The surrounding areas hold Azure-winged Magpie.

About 5km after the bridge, a cross-roads is reached and a narrow road to the right leads to the village of Estrela, which lies very close to the water. Collared Pratincole can sometimes be seen here in spring.

Barrancos

This is a Portuguese village with some Spanish traditions where the patois is a mixture of Spanish and Portuguese and is one of the few places in Portugal where bullfights end with the death of the beast. The surrounding area is relatively undisturbed, as the density of inhabitants is among the lowest in the country.

Lying in such a remote corner, the region offers unparalleled conditions for large soaring birds and there are also many interesting woodland birds.

Location *Geographical coordinates: 38°08′09.30″N, 6°58′35.00″W*

The village of Barrancos lies in the far eastern corner of the southern Alentejo, very close to the Spanish border. Full details of the area can be found on the following maps: 1:100,000: 44; 1:50,000: 44-A, 44-B, 44-CD; 1:25,000: 503, 504, 514, 515 and 525.

Description There are no towns or villages between Amareleja or Safara, in the west, and Barrancos, in the east, with only a few houses scattered throughout the region. A few kilometres north-west of Barrancos lies the old castle of Noudar, an old medieval fortress which is a must for its historical aspects, its landscape and its birds.

The region has not suffered overly from changes wrought by man; it is hilly and wooded, with Holm Oaks the most conspicuous tree species, while Sun Rose is found where woods give way to dense scrub. The Ardila and Murtigão rivers run westwards through the area, before entering the Guadiana, forming unusually narrow, deep and relatively undisturbed valleys. Its value in terms of biodiversity and scarce species can be compared to the much larger International Tagus and Douro valleys. The dryness of the soil and the low population density have produced a kind of 'desertification', although cattle grazing is an important activity in this region and the famous black pigs of the Alentejo can be found here. From the conservation point of view, this area is very important and its bird life is one of the richest in Portugal away from the coast.

Birds

This is a good place for birds of prey and vultures. Griffon Vultures are quite common and small flocks are regularly seen, sometimes with one or two Cinereous Vultures among them. Other common or fairly common breeding raptors are Black Kite, Montagu's Harrier, Short-toed Eagle and Buzzard, while Black-winged Kite is scarce. Red Kite occurs in winter and there are one or two pairs of Bonelli's and Golden Eagles, but these birds are scarce and endangered. Spanish Imperial Eagle is sometimes seen. Black Storks are another speciality of this area and can sometimes be seen flying over the valleys, while the elusive Eagle Owl often calls at dusk from late winter onwards and is sometimes heard in the more remote valleys.

Apart from its large soaring birds, Barrancos is also outstanding for its small landbirds. Common species in the Holm Oak woods include Bee-eater, Thekla Lark, Woodlark, Black-eared Wheatear, Subalpine, Sardinian and Western Orphean Warblers, Short-toed Treecreeper, Woodchat Shrike, Azure-winged Magpie, Spotless Starling and Rock Sparrow. In the deeper rocky valleys there are Red-rumped Swallow, Crag Martin, Blue Rock Thrush, Raven, Hawfinch, and Rock Bunting; in the places with some vegetation thickets along the water bodies, there are good chances of finding Melodious Warbler. The old castle of Noudar has Crag Martin, Red-rumped Swallow and Black Redstart.

Around Amareleja, the landscape is more open and it is possible to find White Stork, Great Spotted Cuckoo and Golden Oriole in late spring and early summer. This area is also one of the best for wintering Cranes in Portugal, which feed in open Holm Oak woods over most of the region.

Accommodation

Barrancos has few options in terms of accommodation and apart from one or two pensions, there is a small hotel in town. In the surrounding areas, there are some rural houses that were converted to tourism. Another possibility is to stay at Moura, which lies 50km to the west of Barrancos, or Évora (110km to the west) and leave from there at dawn.

Strategy

The best time to visit Barrancos is in spring, but several species are resident and can be found throughout the year.

The easiest way to get to this area from Lisbon is to follow the A6 to Évora, then proceed eastwards on the N256 to Mourão and finally south-eastwards to Amareleja.

There are two roads leading to Barrancos, both coming from the west – one from Amareleja (N386) and the other one from Safara (N258), forming a triangle of considerable interest. The following is a description of a circuit starting at Amareleja and ending at Safara, but of course it can be done the other way round.

Begin by exploring the area around Amareleja. Leaving on the N386 eastwards, stop at the cemetery just before a plantation of Umbrella Pines; this is a good place to find Great Spotted Cuckoo, Red-rumped Swallow and Golden Oriole in summer.

Proceeding eastwards on the N386, there is a bridge near km 47.5 over the river Ardila. This site is also worth a stop, as Red-rumped Swallow, Azure-winged Magpie and Hawfinch can all seen from the bridge. From here eastwards, Griffon Vultures are commonly seen. Furthermore, a walk northwards along the river valley may produce Bonelli's Eagles and other birds of prey.

The road continues eastwards, through Holm Oaks. Here it is possible to find the typical species of oak woodland, mentioned above, so it is worth stopping now and then in order to explore near the road.

On arriving at Barrancos, turn left at the petrol station, following a narrow surfaced road that is signposted 'Noudar'. After 3km, stop at the

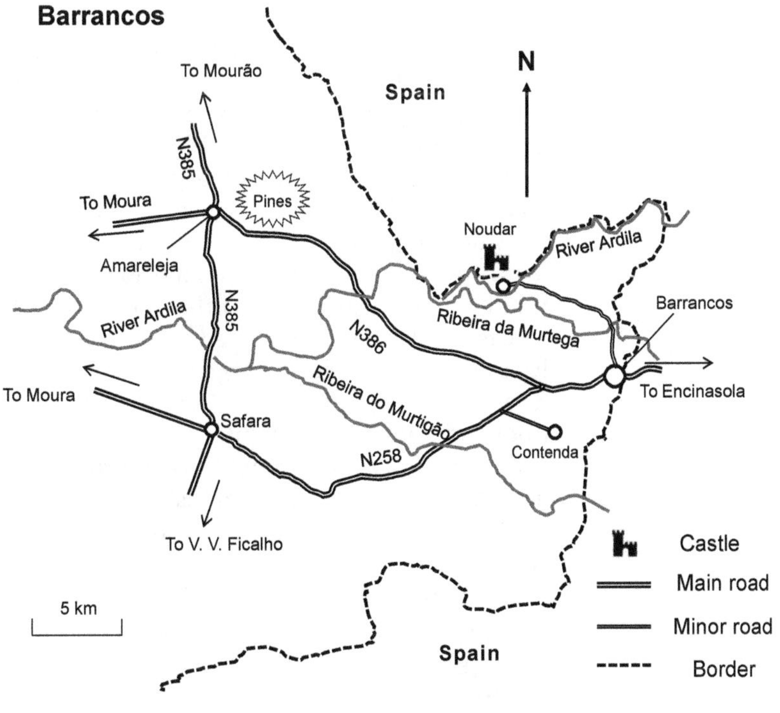

bridge over ribeira da Murtega. Crag Martin, Red-rumped Swallow and Blue Rock Thrush, among others, are all possible. After crossing the bridge, take the left fork and proceed towards the old castle of Noudar. The road is not surfaced, but is in fairly good condition, and leads through extensive Holm Oaks, which hold Western Orphean Warbler – this species is often detected by its loud song, usually heard from mid-April onwards.

After 10km from Barrancos, one arrives at the castle of Noudar, where it is worth spending some time. The tower in the middle of the castle is an excellent watching point for large soaring birds, including Black Stork, large vultures and the occasional Golden and Imperial Eagles. The best time to see them is around noon. The area near the castle is worth a detailed examination and after dusk this is a good site for Eagle Owl, with calling birds being recorded annually.

Back at Barrancos, leave westwards on the N258 towards Safara and Moura and turn left near km 97 to the farm of Contenda. Vultures have survived until recently because they have been regularly fed by the Forestry Service. Griffon Vultures are usually seen in considerable numbers, often with a few Cinereous Vultures among them. Eagle Owl and Western Orphean Warbler also occur here. Further to the west on the N258, the bridge over ribeira do Murtigão is also worth a stop: Black Stork is known to occur here.

Lagoa dos Patos

Geographical coordinates: 38°09′27.70″N, 8°03′57.00″W

A group of two reservoirs of variable extent, fluctuating markedly in dry winters or when used to flood neighbouring rice fields in summer. They lie about 20km north-west of Beja, just south of a much larger dam, barragem de Odivelas. The larger one, called Lagoa dos Patos (literally: 'duck lake') is usually most interesting birdwise than the smaller one, Lagoa do Peneireiro ('Kestrel lake').

Interesting species in the breeding season include Great Crested Grebe, Grey Heron, Gadwall, Black-winged Kite, Montagu's Harrier, Collared Pratincole, Little Ringed Plover, Great Spotted Cuckoo, Short-toed Lark and Black-eared Wheatear. In winter, the lake can hold thousands of wildfowl, mainly Shoveler, Gadwall, Teal and sometimes a few Greylag Geese. Large numbers of Cormorants also occur and Great White Egret is regularly recorded. Black Stork turns up at times. Red Kite is regular here and some Little Bustards can be seen nearby. Passage species include Garganey, waders, Gull-billed Tern and marsh terns, Bee-eater and Tree Pipit. Rare birds have included Red-knobbed Coot and Red-throated Pipit.

Access is relatively straightforward from Ferreira do Alentejo (which lies 20km west of Beja). From Lisbon, take the A2 southwards and leave it after 120km, following signs to Beja. About 5km east of Ferreira do Alentejo (on the N121) turn left at a signpost reading 'Cuba' and 'Peroguarda'. About 3km further on the road passes through Peroguarda village. Look for signs reading 'Alfundão'. In Alfundão, follow the signs to Alvito. About 3.6km later turn left (westwards) onto a sandy track (the track lies opposite a farm on a slight rise with some scattered eucalyptus trees around it). After about 500m, the first reservoir appears ahead.

The best strategy on reaching the reservoir is to drive around the western end and explore the rice fields.

Castro Verde

This extensive area contains the largest expanse of uncultivated plains in southern Portugal. A big part of the Castro Verde region is included in a Special Protection Area (SPA) and efforts are being made to conserve it. This region is especially noteworthy for its steppe birds. For the foreign visitor it offers a unique opportunity to see several very scarce or endangered bird species not only rare in Portugal but also very rare or absent elsewhere in Europe.

Location

Geographical coordinates: 37°42′05.00″N, 8°05′06.00″W

This site lies in southern Portugal, about 50km north of the Algarve. Full details of the area can be found on the following maps: 1:100,000: 45 and 46; 1:50,000: 45-B, 46-A, B; 1:25,000: 539, 549, 549.

Description

This is an impressive steppe-like area, where the rolling plains form a vast ocean of undulating grass and wheat. Cultivated fields alternate with uncultivated ones and due to the dryness of the soil, fields ploughed in one year are left fallow in the following one. Woodlands are almost absent and restricted to a few Holm Oak woods, most of which lie far from the main roads. Some of them have been cleared of the original understorey, once formed by Sun Rose. Here and there, small eucalyptus plantations break the grass-rich landscape and there are some tiny plantations of Olive trees also. Some Umbrella Pine plantations have been established in recent years. The main roads are bordered with large eucalyptus trees.

Water is a scarce resource in this part of Portugal. A few streams flow through the area, but some of them dry out completely in summer. There are also a few small reservoirs, used mainly for agriculture.

As in the rest of inland Portugal, the 'flight from the land' has concentrated the Alentejo's ageing population in the existing villages and small towns. Most isolated houses have been abandoned and, in some cases, are in a state of advanced decay. Almodôvar and Castro Verde are the only 'large' towns in this strange but very beautiful place.

Birds

This is probably the best place in the country for several steppe and open-country bird species.

Great Bustard is one of the most sought-after species and these plains hold a few hundred of them, mainly to the north and east of Castro Verde. Little Bustard is quite common and in spring is found almost everywhere, but in winter in tends to gather in flocks.

Lesser Kestrel and Roller are reasonably common and can frequently be found around old buildings. Finally, Calandra Lark is a very common resident.

Most sightings of Black-bellied Sandgrouse in Portugal are from around this area or around Mourão. However, sightings cannot be guaranteed and it is difficult to recommend a specific site as birds can turn up anywhere. Even so, the eastern part, around Alcaria Ruiva or Penilhos (see below), is most favourable. Stone-curlew can also be found here.

Apart from the above-mentioned species, the plains of Castro Verde are excellent for several other groups of birds. Raptors are reasonably common here and on a day trip in the appropriate season one can expect to find most of the species typical of southern Europe. Spanish Imperial Eagle has recently started to breed and can be regularly seen. There are also one or two pairs of Bonelli's and Golden Eagles. Cinereous and Griffon Vultures do not breed but they are seen frequently in the area. In summer, apart from the already mentioned Lesser Kestrel, several other raptor species put in appearance, namely Black Kite, Montagu's Harrier and Booted and Short-toed Eagles, while in winter one can expect to see Hen Harrier, Red Kite and the odd Merlin. By adding in the three common resident species (Buzzard, Kestrel and Black-winged Kite), the variety of raptor species can exceed ten on a good day.

Rufous Bush Chat

In spring and summer, one can also find a wide variety of interesting migrant/breeding species. Bee-eater is common and can be seen everywhere. Other species which are widespread in this region include Short-toed Lark, Red-rumped Swallow, Black-eared Wheatear, Woodchat Shrike and Golden Oriole. Less numerous species which require a more careful search are Collared Pratincole, Great Spotted Cuckoo, Red-necked Nightjar, Pallid Swift, Tawny Pipit, Melodious, Western Orphean, Spectacled and Subalpine Warblers. The rare Rufous Bush Chat is regular in the eastern part of the area and can be found mainly along rivers and streams, usually from early May onwards.

Other species found on the plains throughout the year are White Stork, Stone-curlew, Thekla Lark and Raven, while the more wooded areas hold

Azure-winged Magpie. Spotless Starling is a common resident throughout the region. Spanish Sparrows are common breeders, often nesting conspicuously under White Storks nests; they also stay in the area during winter but they can be harder to see and identify at this time of year as they are not in breeding plumage and look duller.

Winter brings a few northern migrants to the plains. This is one of the four main sites for wintering Cranes in Portugal (the others being at Évora, Elvas and Mourão) and a few hundred birds can be seen regularly flying over the plains in the evening. Other birds commonly seen during the cold months include Lapwing and Golden Plover (sometimes forming large mixed flocks) – Sociable Plover has been seen here so all Lapwing flocks are worth inspecting. Skylarks are also numerous in winter. Waterbirds are not as numerous and tend to concentrate at the small ponds and reservoirs. Among the species to be found are herons, egrets and a few waders such as Black-winged Stilt, Snipe, Greenshank and Green Sandpiper and sometimes ducks, including Gadwall and Red-crested Pochard.

Accommodation

Accommodation facilities are limited, but there is one hotel in Castro Verde. Alternatively, it is possible to base oneself in the nearby village of Mértola, about half an hour drive to the east, or at Beja, a larger town lying about 40 minutes drive to the north-east. The area is also close enough to the Algarve to allow a day trip from the south coast.

Strategy

These plains are good for birds throughout the year, but winter and spring are certainly the best seasons.

Castro Verde can easily be reached from Lisbon or Faro using the A2 motorway (leave it at exit No. 12 following the signs to 'IP2' and 'Castro Verde').

There are three main roads crossing the area: the N2 which runs north to Aljustrel and south to Almodôvar; the IP2 which leads north-east to Beja and south-west to the A2 and the N123 which goes eastwards towards Mértola. However, all these roads have some traffic so it might be preferable to look for minor side roads, which are quieter and much more suitable for birdwatching.

The whole region is rather uniform, and this means that for many species there are no precise reference points due to the large amount of habitat available. The most interesting birds can appear almost anywhere and this is especially true for steppe birds. It is therefore wise to explore the area as a whole, rather than fixing on any given site. Take more time than you would normally take and wait patiently, particularly towards evening and early morning, preferably near some small pool or minor reservoir. Most good sites are scattered throughout the vast area, making it useful to have a car. While driving, occasionally glance up at the telegraph wires, which are frequently used as perches by Bee-eaters, Black-eared Wheatears, shrikes and sometimes, Rollers.

a) Entradas area

This sector lies to the north-east and can be explored using either the N2 or the IP2.

One of the best ways to start exploring this region is by visiting the Environmental Education Centre of Vale Gonçalinho. This is located between Castro Verde and the village of Entradas. The property is run by Liga para a Protecção da Natureza (LPN) an environmental NGO with much good work to its credit. Besides general information, there are several foot tracks nearby that can be followed, allowing a good introduction to the birds of the area (the centre is open Tuesday to Saturday from 9 am to 6 pm). For the keen photographers there is the possibility of hiring a hide to try photographing Great Bustards and other steppe birds. Reaching Vale Gonçalinho is simple. From Castro Verde just follow the signs starting at the roundabout close to the petrol station. These will lead you to the old abandoned N391 road which runs parallel to the IP2 going towards Beja. Immediately after km 6 veer right where signposted 'LPN – Centro de Educação Ambiental Herdade de Vale Gonçalinho'. Roads may be confusing here so, if for some reason you end up on the IP2, you should leave it as soon as possible to the N391 (Vale Gonçalinho turn is between km 382 and 383 of the IP2).

Back to the main road and proceeding north-eastwards, after a few more km, one reaches the village of Entradas. From here, there are two roads worth exploring: one that runs to the north-west and the other one leading south-east.

The tarmac road to the north-west leads to Carregueiro and it crosses good habitat for steppe birds. Along the 9.5km of this road there are a few rough tracks leading to 'montes' (rural houses). Some of these tracks are open and are worth exploring. For instance the track that starts some 3.2km after Entradas signposted 'Monte Branco' will lead to a geodetic mark after 1.6km from where it is possible to have good views over the region and see Great Bustard and Crane. Some of these rough tracks get very muddy in wet winters and some may not be passable with a normal car at all. Also note that the tarmac road is quite narrow here and although there is little traffic, extra care should be taken when stopping along here.

To take the other road from Entradas, look for a tall water deposit in the village. From there, a minor road which is not signposted leads south-east towards São Marcos de Ataboeira (10km away). Along its first half the road crosses open country, good for steppe birds. The second half passes through Holm Oak and Cork Oak, good for passerines. About 8km after Entradas the road crosses the small ribeira de Cobres. It is worth stopping here and taking a walk along its banks as this is a good site in late spring and early summer for Rufous Bush Chat – an uncommon species in Portugal. After a few more km, upon reaching the village of São Marcos da Ataboeira, you reach the N123 which runs from Castro Verde to Mértola (see following section).

b) Mértola road

The key to exploring this region is the N123, which runs from Castro Verde eastwards to Mértola. Most birds can be seen along the road or within viewing distance of it. Be very careful when stopping along this road and where possible pull off the tarmac.

After a few more km the village of São Marcos da Ataboeira appears on the left. From here a minor road on the left (not signposted) leads to Entradas and is worth exploring, as detailed above.

Close to km 77 there is a turn to the left signposted 'Salto' and 'Senhora de Araceli'. This leads one along a tarmac but narrow road to the small chapel of Araceli (or Aracelis), after 9km. This beautiful church (which can also be reached from the turn between km 83 and 84 on the N123) is located on a small hill giving excellent views of the neighbouring region. The cultivated areas around are very good for steppe birds. The chapel itself is an excellent viewing point to look for soaring raptors, including vultures, kites and eagles.

Proceeding eastwards on the N123 a hill will appear on the right hand side with some conspicuous radio antennas on top. This is pico de Alcaria Ruiva which rises to 370m. Access is by a very rough and sloping track that can be taken, turning right just after km 92 (high clearance vehicle or four wheel drive advisable). The effort of actually climbing to the top is rewarding with outstanding views and the possibility of watching large raptors including Spanish Imperial Eagle.

To the south of the N123 lies a network of minor roads (linking the N123 to the N267, which runs further south and links Almodôvar to Mértola) which are also worth exploring. A possibility is to draw a circuit through Viseus, Figueirinha, Penilhos and João Serra, stopping wherever possible to scan for birds. This area holds Black-bellied Sandgrouse, but this species is not always easy to find – try to listen for its characteristic calls. Spanish Imperial Eagle can sometimes be seen here. It is recommended that you walk, rather than just drive through the plains, especially over uncultivated fields in the very early morning.

c) Casével area

The village of Casével lies about 10km west of Castro Verde and can

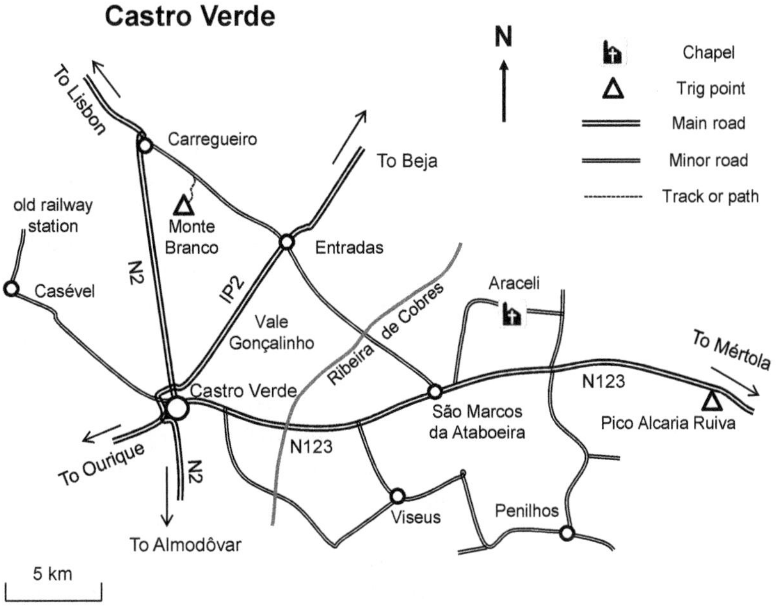

be reached using the municipal road M535, which leaves from the 'sheep roundabout' located near the centre of Castro Verde and runs past the school and the local stadium. It crosses a mix of open habitats, with some Olive groves. This road has less traffic than the N123 described above, so it is less disturbed. Along this route it is possible to see several species of passerines and also some birds of prey.

Close to the road there are a few ponds, which are worth inspecting, as they often hold some waterbirds – depending on season, one can expect to find grebes, egrets, herons, ducks, coots or waders.

At Casével it is worth proceeding northwards along a narrow quiet road which takes you to the old railway station. Look for signs reading 'Estação de Casével'.

Mértola

Geographical coordinates: 37°38'33.50"N, 7°39'34.30"W

Mértola is a small town built on the right bank of the Guadiana river, in the extreme south-east of Baixo Alentejo, not far from the Algarve. It has a very rich history that goes back to Portugal's Islamic period. Apart from being scenically unique, it is very interesting from the birding point of view as well. The same applies to the surrounding area. A large part of this region (about 69,700ha) was included in a natural park created in 1995.

The river Guadiana runs just past Mértola, but its valley is very deep and dry here and is rather devoid of aquatic vegetation. However, in the narrow valleys of the smaller tributaries the vegetation is richer; Sedge and Oleander abound, and there are often a few other trees like Ashes and Poplars. East of Mértola, the soil is even drier, the area is semi-desert in biogeographical terms and the cultivated fields are restricted. Sun Rose is the dominant wild species of plant and forms large monotonous patches. At Mina de São Domingos there are extensive ruins of an old abandoned copper mine, nowadays full of leached sulphate deposits, with many Oleanders and eucalyptus. The whole place has a character unlike any other in Portugal, part Mediterranean scrubland, part multicoloured lunar landscape.

Some of the steppe birds that can be seen at the neighbouring site of Castro Verde can also be found here, although in smaller numbers, such as Lesser Kestrel, Great Bustard, Little Bustard and Black-bellied Sandgrouse.

Large raptors do also occur and among them Cinereous and Griffon Vultures are regular visitors from nearby Spain and Golden, Spanish Imperial Eagle and Bonelli's Eagles may breed nearby. Eagle Owl also breeds in the area and there are a few pairs of Black Stork.

The river valleys and the scrubby areas are very good for smaller birds and Crag Martin, Red-rumped Swallow, Blue Rock Thrush, Golden Oriole, Azure-winged Magpie and Rock Bunting are not uncommon. Other summer visitors in this region include Black-eared Wheatear, Subalpine, Spectacled and Western Orphean Warblers and Woodchat Shrike. Rufous Bush Chat also occurs and some of the very few Portuguese records of Western Olivaceous Warbler were made in this region. Mértola area is probably the most reliable place in the country to find White-rumped Swift, a very rare and localised breeder. Thekla Lark is a common resident.

If coming from south (Algarve), the best way is to take the N122 that runs from Vila Real de Santo António and Castro Marim, in the east, straight to Mértola or to take the A2 northwards to Gomes Aires (45km) and then the N393 eastwards through Almodôvar to Mértola. From Lisbon access is also via the A2 to Grândola and then proceed on the IP8 through Beja and finally the N122 southwards.

Exploration of this area can start at Mértola itself which holds a colony of Lesser Kestrels. Birds breed in the Convento de São Francisco just outside the town and are carefully monitored and protected by the owners. It is necessary to ask for permission to visit the old convent and to watch the falcons. Disturbance of any kind is not tolerated. Other species breeding at Mértola include White Stork, Blue Rock Thrush and Golden Oriole.

About 20km north of Mértola, there is a place called Pulo do Lobo ('jump of the wolf'). In that place the river Guadiana almost disappears in a very narrow and rocky gorge that, according to legend, in the old times could be jumped by a wolf. It is a beautiful place that surely deserves a visit, not only for the geological and scenic aspects but also because of its birds. Black Stork and Bonelli's Eagles can sometimes be seen. Access to Pulo do Lobo is signposted and can be made from the N122 a few kilometres north-west of Mértola, following signs to 'Corte Gafo' and 'Amendoeira'.

The left bank of river Guadiana can be very interesting from the ornithological point of view as well. Crossing the river at Mértola and driving 4km to the east on the N265, turn right onto a minor road that leads past Fernandes and Tamejoso. This area is good for Spectacled Warbler – a rare breeding species in Portugal – look for it in low scrub. Further east lies the Mina de São Domingos, best approached from the tiny village of Santana de Cambas. From this village, head north along rough stony trails, through Sun Rose scrub (a good habitat for Subalpine Warbler), towards the chimneys of the old mine, just visible in the distance. White-rumped Swifts are regular there in summer and have bred. In this area, there are often sightings of Black Stork, Griffon and Cinereous Vultures.

ALGARVE

Sagres and Cabo de São Vicente

The myth of Sagres as a school of navigation played an important role in European history although much of the true history surrounding its importance in Portuguese discoveries has recently been revised. Nevertheless, voyages from Portugal to Africa and Asia, which had begun from the area in the 14th century, were thoroughly organised by the 15th century, aided by instruments acquired from the Moorish invaders of North Africa. Sagres, close to Europe's lands-end became a byword in navigation circles, aided in part by the interest of King John I and his son's patronage; the son became known as Henry the Navigator.

From the ornithological point of view today its strategic location makes it an extremely important migration watchpoint, especially in autumn. Because it forms the south-westernmost corner of Europe, unexpected birds can and do appear here, amid the myriad of more common species. The place has a mystical atmosphere – a happy conjunction of history, geology, climate and wildlife.

Location *Geographical coordinates: 37°00′28.00″N, 8°56′41.00″W*

The headlands at Sagres and Cabo de São Vicente (Cape St. Vincent) are situated at the south-west tip of Portugal, in the western Algarve. Full details of the area can be found on the following maps: 1:100,000: 51; 1:50,000: 51-B; 1:25,000: 601 and 609.

Description Cabo de São Vicente itself is located where the Portuguese southern and western coasts meet. It forms therefore the very sharp 'corner' of Europe on its southern flank.

The coastline is mainly formed by rocky sea cliffs, with small bays and beaches between. On the western coast, the wind is stronger and the sea can be quite wild, while the southern shore is markedly protected from wind and cold.

The whole area is very dry, the landscape being rocky with low bushy vegetation. Rare plant species, some of them endemic with an extremely limited geographical range, can be found here and should not be disturbed or removed. Wheat is the main cereal cultivated, while in uncultivated fields the main activity is sheep grazing, producing a 'downs' effect, not unlike southern England. There are small woods of Maritime Pine dotted about, otherwise the area is almost treeless.

Human presence is marked, with a lighthouse at the cape (visible at long range), a light beacon at Sagres within the fortress walls and a harbour village, Sagres itself. There are also some scattered farms. Tourism is growing here, but as yet has not disfigured the landscape as has happened further east along the coast.

Birds In autumn, large numbers of migrants concentrate in this area. Though much less spectacular a phenomenon than at other famous places, like the Strait of Gibraltar, the birds are often numerous and certainly most Iberian and west European soaring birds can be seen without difficulty.

Booted Eagle is the most common soaring species and on good days, one may see 30 or 40 birds; flocks of 400 have been recorded here. Other common species are Honey-buzzard, Black Kite, Egyptian Vulture, Short-toed Eagle, and Sparrowhawk. Flocks of Griffon Vultures, sometimes with hundreds of birds, occur, although they tend to appear late in the season. Less common species include Cinereous Vulture, Golden Eagle, Spanish Imperial Eagle, Red Kite, Lesser Kestrel, Hobby and Eleonora's Falcon, and rarities have included Rueppell's Vulture, Pallid Harrier, Lesser Spotted Eagle and Red-footed Falcon. Bonelli's Eagle and Peregrine Falcon occur regularly, although most sightings probably refer to resident birds breeding nearby. Strangely, Black Stork is more regular than White Stork, unlike the situation found at Gibraltar.

The migration of passerines is also very interesting and, if weather conditions are favourable, some species may turn up in good numbers, especially in the bushy valleys and in the pinewoods. The commonest ones are Nightingale, Redstart, Whitethroat, Garden and Willow Warblers, flycatchers and Woodchat Shrike. Less common but still regular are Wryneck, Spectacled (which also breeds), Subalpine and Western Bonelli's Warblers and Ortolan Bunting. The more open areas often hold migrating Tawny Pipits, Short-toed Larks and several subspecies of Yellow Wagtail, plus Whinchat and Wheatear. Dotterel also occurs at times, being best looked for in uncultivated, dry fields.

Spring migration is markedly weak by comparison, as migrants move through eastern and central Iberia where passage is rapid and purposeful. Trans-Saharan migrants tend to leapfrog over Iberia and may miss the Algarve altogether. 'Fall' conditions and easterly winds are necessary to see much movement of long range migrants and these conditions may only occur once or twice in an Iberian spring, mainly in April or early May (but in some years these conditions may not occur at all).

Migration of seabirds is also noteworthy. In autumn huge numbers of Gannets can be seen passing south. Cory's Shearwater is a common sight, sometimes forming large 'rafts' on the sea inshore. Other common birds are Balearic Shearwater and Great Skua. Great and Sooty Shearwaters and Pomarine Skua are regular.

The variety of resident birds is small. In the dry areas the most interesting species include Little Bustard, Thekla Lark, Fan-tailed, Sardinian and Dartford Warblers, Chough and Spotless Starling. On the rocky coast, Shag, Peregrine Falcon, Yellow-legged Gull and Blue Rock Thrush are the most noteworthy. All these species can easily be found. In spring and summer Pallid and Alpine Swifts are not uncommon and Spectacled Warbler occurs in scrubby areas. In winter, there are good numbers of Lapwing, Golden Plover and a few Stone-curlews. Richard's Pipit is regularly recorded in small numbers. Alpine Accentor

Booted Eagle

sometimes occurs at the rocky bluffs just below the lighthouse (care is needed in exploring the cliff edges!).

Several rarities have also been recorded in the area, either in the beaches or at the harbour. Among them are: Lesser Yellowlegs, Iceland, Glaucous, Laughing and Sabine's Gulls and Lesser Crested Tern have turned up at the harbour. Snow Finch has been seen near the lighthouse.

Accommodation

This area is an acknowledged tourist attraction and accommodation should not be a problem, except possibly in mid-summer. At Sagres, there is a wide range of possibilities, from the more simple pensions to expensive hotels. There is also a camping site at Sagres. The area is only half an hour drive from Lagos which has a good variety of accommodation options.

Strategy

Autumn (mid-August to October) is by far the best season to visit this place. At that time of year, the post-breeding migration is at its peak. For soaring birds, it is best to choose a point from where to watch most of the area and wait for the birds to fly over, mainly between 10 am and 1 pm (this is the time when thermal currents start forming as the temperature rises). One of the best points is located near a pinewood, north-west of Sagres village. To get there, leave Sagres on the N268 that leads westwards to the Cabo de São Vicente, turning to the right at the curve which lies after the Aldeamentos São Vicente. Continue along a rough track, which follows a pine hedge (good for passerines at times), turning right at the end and proceed to the geodetic mark (marked 'Cabranosa' on 1:25,000 maps). At Sagres, the passage of birds of prey can either be spectacular or frustrating and the number of birds present can change markedly from one day to the other, according to the weather conditions. When the weather is anticyclonic with easterly or north-easterly winds, large numbers of birds of prey concentrate in the area. Ringing has indicated that some of these birds may stay around for several weeks, probably because, unlike Gibraltar, Cabo de São Vicente truly is lands-end. When the wind has a more westerly tendency, raptor passage is minimal.

Apart from the soaring birds, passerines occur close to the coastline or in the pinewoods. A visit to the valley close to Telheiro beach, which is located about 2 or 3km north of the cape, could also be rewarding.

Between the cape itself and the more extensive pinewoods to the north, the ground is open and dry and several interesting species can be found. To get there, turn right on the second road after Aldeamentos São Vicente and proceed along a narrow tarmac road until a small farm is reached. This is called Vale Santo and is a good spot for Chough, Tawny Pipit and Yellow Wagtail. Dotterel has been seen on several occasions in the flat areas near the farm and there is usually a small flock of Little Bustards around.

Sagres harbour and the small ephemeral lagoon at Martinhal beach, north-east of Sagres, are also places to visit. Usually there are not many birds there apart from gulls at the harbour and small flocks of waders at the lagoon especially in autumn. However, both sites have great potential and should not be neglected and to prove it there are some fascinating records of rarities at both places.

Seawatching can be very rewarding especially in southerly winds. At Sagres try either the fortress or Ponta da Atalaia (not far from the pousada). Both are within walking distance from the village centre. The best spot in the area however is Cabo de São Vicente. Note that this headland is very high and that it is very dangerous to try to descend further because of the scree on the slopes of the cliff. The best strategy here is to find a good and safe place to sit and wait (care should be taken with the wind, which can be strong here). Watch out for seabirds passing very close, which fly almost beneath the observer at times. Early morning is best as bird movement tends to be higher during this time of day and because the sun will be at your back.

This area, the western approaches to the Gulf of Cadiz, is very rich at times and its proximity to Macaronesian waters and the African coast makes any seatrip exciting. There are now some companies that will organise pelagic trips to carry out birdwatching but another option is to just get a seat on a day-long fishing cruise. These trips can be very productive, especially in autumn when there are high chances of seeing Wilson's Petrel (sometimes in moderate numbers and well within camera range) and Sabine's Gull.

In the past the Instituto da Conservação da Natureza e das Florestas (ICNF, formely ICN) and Sociedade Portuguesa para o Estudo das Aves (SPEA) have sponsored autumn raptor counts from August to October, mainly from 1990 to 2001. That is not the case anymore although SPEA now organises an annual autumn bird festival.

Sagres can be reached from Faro using the A22 until Lagos, then the N125 to Vila do Bispo and finally southwards on the N268.

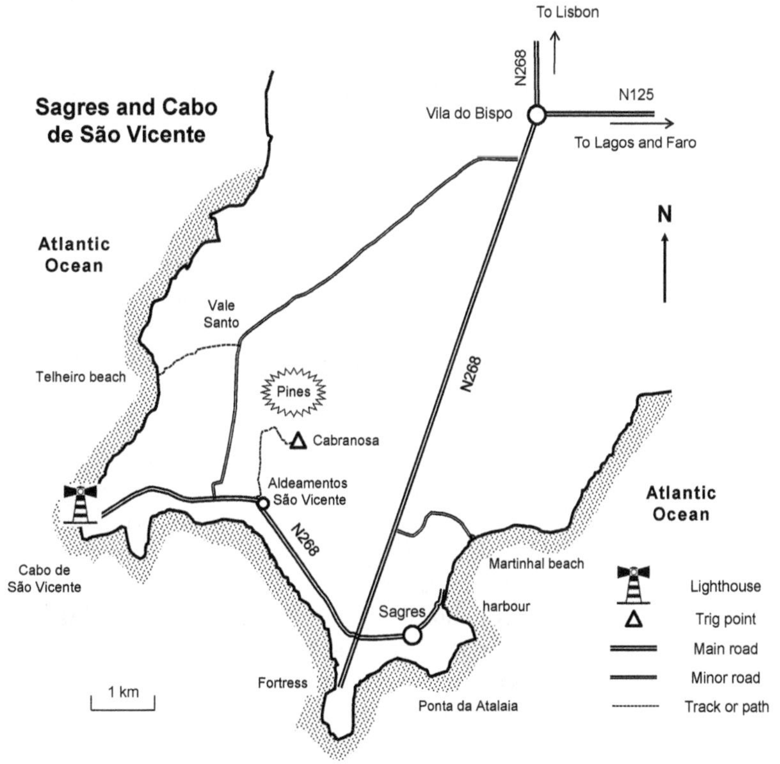

Ria de Alvor

Geographical coordinates: 37°07′56.80″N, 8°36′43.30″W

Ria de Alvor is a small estuarine wetland located between Portimão and Lagos, in the western Algarve. It holds a variety of habitats, from saltmarsh to bushy areas, together with open fields, pinewoods, almond orchards and market gardens.

Breeding species include Black-winged Stilt, Stone-curlew, Kentish Plover, Little Tern, Hoopoe, Bee-eater, Red-necked Nightjar, Crested Lark, Melodious Warbler, Woodchat Shrike and Serin, while Cattle and Little Egrets breed nearby and usually feed in the area. Passage can be exciting, especially in September and October, as the southern coast of Portugal is a stopping off point for trans-Saharan migrants. At that time, it is possible to see many swifts, pipits, hirundines, chats, warblers and flycatchers. Passage of waders is also interesting, with most west European species present, including Curlew Sandpiper, Greenshank, Ruff, Knot, Whimbrel and both godwits; gulls and terns turn up too and Black Terns are regularly seen, while less common species include Audouin's Gull (mainly in August). In winter, a few specialities can be seen, including Greater Flamingo and Spoonbill; Caspian Tern is regular, as are several waders and some interesting passerines including Bluethroat, Firecrest, Southern Grey Shrike and sometimes Penduline Tit. Osprey and Marsh Harriers are regular non-breeding visitors.

To get there from Faro, take the A22 motorway westwards. After about 65km, leave the motorway traveling on node number 4. From there follow N125 westwards for about 4km, turning left almost precisely at the km 35 marker, opposite a signpost to the right reading Mexilhoeira Grande. By crossing the railway line (take care!) and following the unsurfaced track for about 2.5km through almond orchards, the estuarine area can be reached. Park the car and walk.

Serra de Monchique

Geographical coordinates: 37°19′05.00″N, 8°33′21.20″W

Away from the coast, the Algarve turns hilly and greener, in sharp contrast to the flat lowlands closer to the sea.

Peaking at 903m, serra de Monchique is the highest point in the Algarve. For the birdwatcher, this area is worth visiting, as it holds a number of birds that are rare along the coast. Above all, it is good for forest birds – here one can expect to find woodpeckers, tits, Woodlark, Short-toed Treecreeper and Cirl Bunting. Birds around the top typically include Woodlark, Red-rumped-Swallow, Dartford Warbler, Linnet and Rock Bunting; passage sometimes brings a few surprises, like Rock Thrush and Ring Ouzel. Raptors are likely to be seen in the area, including Bonelli's and Short-toed Eagles.

Monchique town lies in the western Algarve, about 25km north of Portimão. If coming from Faro or Lagos, follow the A22 motorway and leave it when reaching the exit to Portimão, but following the signs to Monchique instead.

The road is winding and goes uphill; after about 15km one reaches

Caldas de Monchique (the spa area) where there is a car park; it is possible to move around on foot from here. Some garden and forest birds can be found.

Proceeding on the N266 for another 7km one eventually reaches Monchique town. From here, it is worth proceeding to the peak named Fóia which can be easily reached on a tarmac road. The area around the top is virtually treeless. On clear mornings, the view is breath-taking.

From Monchique, there are other roads worth exploring, like the N266 northwards (which will bring you through Cork Oak woodland) or the N267 westwards from Nave (3km south of Monchique) than runs to Marmelete. Bonelli's Eagle is possible here.

Lagoa dos Salgados

This small coastal lagoon remained relatively unknown until the mid-1990s, when it was put on the map thanks to the efforts of some local birdwatchers and soon became one of the Algarve's birdwatching hotspots.

Location

Geographical coordinates: 37°05′44.00″N, 8°20′10.00″W

The site lies right in the heart of the tourists' Algarve, between the cosmopolitan town of Albufeira and the crowded village of Armação de Pera. Maps for this site are: 1:100,000: 52; 1:50,000: 52-B; 1:25,000: 604.

Description

The Lagoa dos Salgados occupies about 150ha. It is formed by a main body of water at the confluence of two small rivers and by a vast flooded area that extends to the north-west. The lagoon is separated from the sea by a barrier of sand dunes and only very occasionally, either due to heavy rain conditions or human intervention, is this chain broken, allowing sea water to invade.

On the north and west banks there are some agricultural fields and some typical Algarve farmland landscape with dry orchards of Fig, Carob and Olive trees along with small vineyards. This area is potentially very interesting though sadly there are plans to build some apartment blocks there. Close to Armação de Pera, the ribeira de Alcantarilha reaches the sea and there are small areas of saltmarsh vegetation close to the river banks. On the east bank of the lagoon there is a golf course and some areas of tamarisk and aquatic vegetation.

The entire region suffers from heavy tourist demand and huge urban complexes have been built on both the eastern part (Salgados and Albufeira) and the western side (Armação de Pera). Praia dos Salgados and Praia Grande are two very well-known bathing beaches nearby. In some years, as happened in 2012, the lagoon can be almost dry in the summer months.

Birds

The list of birds recorded at this place is long and includes mainly waterbirds but also many woodland and farmland species that have been seen in the surrounding areas. Greater Flamingos are present all year round, numbers peaking from late summer through to the winter. There was even a recent failed breeding attempt. If successful this would probably have been the first for the country. Spoonbills and

Glossy Ibises are also seen regularly, similar to Purple Herons and Little Bittern, which are summer visitors. The mudflats on the banks can hold parties of waders, especially at migration times and during winter. Avocet, Dunlin, Little Stint, Black-tailed Godwit, Ruff, and Ringed Plover are not uncommon. It is worth checking these flocks of shorebirds, as rarities like Long-billed Dowitcher, Lesser Yellowlegs and American and Pacific Golden Plovers have already been recorded there. Black-winged Stilt is resident and can easily be found all year round. Several duck species occur and this lagoon is the only place in Portugal where the nationally threatened Ferruginous Duck has been reported breeding in the last few years. Rarities like Lesser Scaup and White-headed Duck have also turned up there. Caspian Tern occurs in winter. Purple Swamphen is common in the marshy vegetation areas.

On the surrounding fields it is possible to find a good selection of typical southern European birds. These include species like Bee Eater, Hoopoe and Woodchat Shrike, as well as many passage migrants, including chats and warblers. Little Owl is not uncommon and Great Spotted Cuckoo sometimes puts in appearance.

Accommodation

Accommodation is not a problem and can be found in a variety of places nearby namely at the cosmopolitan town of Albufeira, but prices can be high during summer months.

Strategy

Lagoa dos Salgados is interesting at all times of the year, as it holds an interesting set of both breeding and non-breeding birds.

Given its central location, this site can be easily reached from any other place in the Algarve in one hour drive or less via the A22 motorway (leave it following the signs to Algoz and Pera and then follow the N125 eastwards up to Alcantarilha). Best access to the area is made either from Salgados beach on the eastern side (close to Albufeira) or Praia Grande beach on the western side. There is a path connecting these two beaches that runs along the southern edge of the lagoon and allows good views of the waters and the dunes. Purple Swamphens and Purple Herons can be seen from here.

The western bank is probably the best option to birdwatch. There is a track that leads to Praia Grande that runs parallel to the bank and allows easy access to the area. From there it is possible to reach the edge of the lagoon and obtain good (sometimes close) views not only of the waders that feed on the muddy parts but also of the flocks of Greater Flamingos that are usually around. When the sky is clear, which is the most common situation, observing from this side is better in the afternoon, when the sun is behind the observer. There are plans to build two observatories here.

Access to Praia Grande is made from the roundabout close to Pera village on the road that connects Albufeira to Armação de Pera. To reach it from Armação de Pera side veer eastwards at the roundabout near the campsite which lies between Alcantarilha and Armação de Pera; proceed eastwards through the next roundabout and when meeting a second roundabout turn right, following the narrow road that leads to the beach and to the western side of the lagoon. If coming from Albufeira, leave this town westwards through Avenida dos Descobrimentos and proceed on the road to Armação de Pera; about 3 km after passing Vale da Parra the above-mentioned roundabout is reached.

From the dirt track that leads to Praia Grande it is also possible to explore the farmland to the west. This can be done either on foot or driving along the rough track that leads westwards towards Armação de Pera. The orchards and scrubs are very good for passerines and other terrestrial birds. Post-breeding migration is one of the best times to visit this place and sometimes there are big numbers of chats and warblers passing through.

Not far from the lagoon, the mouth of the ribeira de Alcantarilha close to Armação de Pera is also a good place to explore. Usually there are some waders around, especially at migration times. This is a very popular area and can be crowded in summer. Find your way to the beach through the town and then park at the back of the football pitch close to the beach.

As is the case with any other tourist-dense places, there have been cases of robbery. So, be sensible; if you walk away from your car do not leave any valuables at sight.

Lagoa dos Salgados

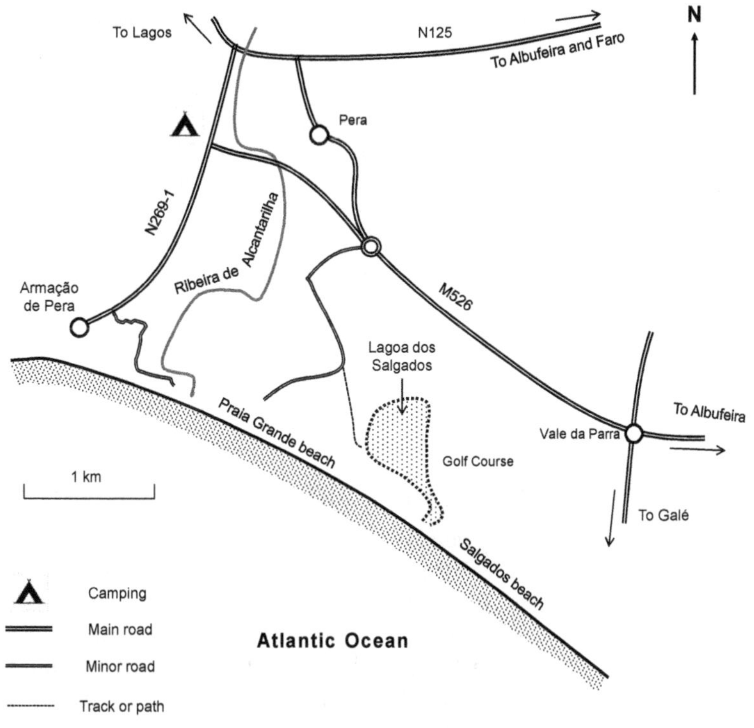

Vilamoura and Vale do Lobo

Geographical coordinates (Vilamoura): 37°05'03.10"N, 8°07'38.40"W

Vilamoura is one of the most popular resorts in the Algarve. There are a few small reedbed patches in the area, namely on the west side, where it possible to find several species typical of this kind of habitat, including Purple Swamphen, Marsh Harrier, Cetti's Warbler and wintering Penduline Tits, along with several species of the more common waterbirds. Additionally, Vilamoura is a good place in which to find Azure-winged Magpies.

The best access to explore it is by following the Avenida Praia da Falésia that leads to the Falésia beach and stopping just before the bridge, which allows a good view over the reedbeds.

Further to the west is the so-called 'Parque Ambiental de Vilamoura'. This is another good spot for waterfowl and a few uncommon species have been seen here. To get there from Vilamoura, drive westwards from roundabout No. 3 on Avenida João Meireles (the main avenue) following the signs to 'Estrada de Albufeira'. After about 2.7km, at roundabout No. 2, turn left and you will find the interpretation centre. About 400m from there find the entrance to the *parque* and a footpath can be taken to explore the area. Alternatively you can proceed on Estrada de Albufeira for another 1.5km and on reaching the Cegonha Country Club turn left and follow the signs to the Parque Ambiental. After about 3km, park and walk – there are two hides near the lagoon.

About 5km south-east of Vilamoura lies Vale do Lobo, another well-known resort. Here there are two small coastal lagoons (Garrão and Dunas Douradas). Both are worth visiting, mainly for their waterbirds which often allow close views. Species to be found here include Purple Swamphen, Little Bittern, various ducks including Ferruginous and sometimes waders on passage. Glossy Ibis has been recorded and Azure-winged Magpies are common in the surrounding pinewoods. To get to this area, leave Vilamoura eastwards, drive through Quarteira and proceed on the road to Almancil, turning right after 4km where signposts read 'Vale do Lobo'. On reaching the big roundabout, turn left. The Lagoa das Dunas Douradas ('golden dunes lagoon') lies close to the Garrão beach (follow signs to 'Garrão – poente'), close to the Dunas Douradas Beach Club. There is a parking nearby and an observation tower. Lagoa do Garrão lies close to the Ancão beach, about 1km further east. This lagoon is concealed among the pinewoods, so the best approach is to park close to the beach and walk from there.

Serra do Caldeirão

Geographical coordinates (Barranco do Velho): 37°14'11.00"N, 7°56'17.00"W

Contrasting with the busy and cosmopolitan coast, the inner Algarve is sparsely populated. The border region between the Algarve and Alentejo provinces is rugged and wild. A barrier of serras stretches along a west-east axis and is dominated by Caldeirão with its 589m height.

The landscape is formed by vast expanses of scrub (mainly Sun Rose) and Cork Oaks but also woods of exotic eucalyptus and plantations of Umbrella Pines. Water is a scarce resource here and the few rivers that flow in the region are often dry through the summer months. Severe fires hit Caldeirão over the years. The last big one was in 2012 and destroyed a vast area.

Bird wise this is a good place to look for Bonelli's Eagle which has a good tree breeding population there. Eagle Owl is not uncommon but as happens in many other places it is difficult to see. Other interesting species can be found here, including Short-toed Eagle, Scops Owl, Red-necked Nightjar, Wryneck, Lesser Spotted and Iberian Green Woodpeckers , Woodlark, Black-eared Wheatear, Blue Rock Thrush, Subalpine, Western Bonelli's and Western Orphean Warblers, Iberian Chiffchaff, Golden Oriole and Rock Bunting.

This is a huge place and exploration is not straightforward. The best option is to use the N2 that starts at Faro and head northwards past São Brás de Alportel and then into the Caldeirão range. About 35km from Faro the N2 reaches the small village of Barranco do Velho. This village lies at a crossroad and from there it is possible to decide the best approach to explore the area.

To the north, the N2 will take you to the Caldeirão summit. From there the views are excellent and an eye should be kept on the sky as sometimes there are raptors around. The scrubby areas nearby are good for some passerines.

To the east of Barranco do Velho and around the village itself there are good areas of Cork Oak woods and Mediterranean scrub that can be explored on foot and where it is possible to find a large number of small birds, namely warblers.

Probably the most interesting place lies to the west, namely around Fonte da Benémola and Rocha da Pena – two small designated sites. The valley of the ribeira da Menalva is very interesting especially the site called Fonte da Benémola. This is one of the very few rivers that does not dry out in summer and it works almost like an oasis. The agricultural landscape attracts a large variety of small birds including several species of warblers. Scops and Eagle Owls occur in this area, the former being detected in hot summer nights by its monotonous call recalling a Midwife Toad. To get here from Barranco do Velho, follow the R124 westwards and then the N396 southwards towards Loulé. About 9km before Loulé look for signs to 'Querença', 'Tôr' and 'Fonte da Benémola'. Proceed towards Tôr until Vale de Mulher is reached. Immediately after this small village there is a sign to Fonte da Benémola.

Rocha da Pena is another well-known place that deserves a visit. This is a limestone escarpment in the edge of the Caldeirão itself. It is an interesting place geologically and it also holds a nice community of cliff-nesting birds.

This place can be reached from the small village of Pena. Best access is from the R124 towards Silves that can be taken a few kilometres south of Barranco do Velho. After about 5km from Salir one must follow the signs to 'Rocha da Pena'.

Ria Formosa

The Ria Formosa is the largest wetland in southern Portugal. It extends for some 60km from Tavira in the east to Ancão in the west and most of this area has been designated a natural park.

Algarve has a very pleasant climate and continues to be visited, more and more frequently, by package tourists. The successive construction of large tourist complexes and all sorts of infrastructures and facilities has led to the destruction of several wetlands and the Ria Formosa, already peppered with illegal constructions and subject to increased disturbance, has become ever more vulnerable.

Nevertheless, 'formosa' meaning beautiful still applies to this extraordinary place, which holds large numbers of breeding, passage and wintering waterbirds.

Location *Geographical coordinates (Ludo): 37°01'54.90"N, 7°59'41.20"W*

Ria Formosa lies immediately adjacent to Faro, Algarve's main town. Full details of the area can be found on the following maps: 1:100,000: 53; 1:50,000: 53-A, B; 1:25,000: 606, 608, 610, 611 and 612.

Description Like other coastal wetlands in Portugal, the Ria Formosa is directly connected to the ocean, but there are a few low islands and islets between the mainland and the sea, which form an almost continuous natural chain of dunes and protect the whole area from the destructive force of the sea. In these dunes, the existing vegetation consists mainly of small bushes and a few pine trees. Over the years some small human settlements were built there. The first ones were made by local fishermen but soon some areas were invaded by illegal constructions made as second homes for tourist purposes. These small hamlets caused serious damages in some stretches of the dunes barrier and a very controversial discussion about demolition has been going for a long time.

Ria Formosa is not a true estuary but even so it is subject to the tidal influence and at the low tide there are extensive areas of mud flats. In the wider sections there are many channels which form an almost labyrinthine net.

Saltpans are a characteristic feature of most coastal wetlands in Portugal and some of those in the Ria Formosa are still actively used for salt production. However, many of them have also been converted to fish farms. From east to west, small streams run into the Ria Formosa, forming coastal pools which are important for birds in season. Some of these pools, surrounded with reeds, are of particular interest for the birdwatchers.

This is a tourist area and the extensive sand beach front is very popular especially in summer months. Besides this, human activities include salt extraction, fishing and shellfish production and collection.

Birds Spring migration begins early, with the first hirundines, Yellow Wagtails and Great Spotted Cuckoos being seen in January.

Breeding birds include Purple Swamphen, Purple Heron, Little Bittern and Audouin's Gull. The coastal islands, mentioned above, hold a large population of Little Terns, as well as a few pairs of Stone-curlews and Short-toed Larks. Other breeding species include White Stork, Collared Pratincole, Sardinian Warbler, Woodchat Shrike, Golden Oriole and Serin.

Late summer and autumn are marked by the variety and abundance of migrant passerines, with Melodious Warblers passing as early as late July; Subalpine and Willow Warblers are seen from mid-August onwards, Western Bonelli's Warbler being recorded on occasion. September is the best month for variety and species to be found usually include Whinchat, Redstart, Bluethroat, Tawny Pipit, Pied Flycatcher, Whitethroat and Grasshopper, Sedge, Reed and Garden Warblers.

Waders are also plentiful at this time of year, with Dunlin, Little Stint, Curlew Sandpiper, Ringed Plover, Ruff, Redshank and Spotted Redshank all common; the odd Green Sandpiper also appears. Birds of prey pass through the area in small numbers, with regular sightings of Booted and Short-toed Eagles, Osprey and occasionally Griffon Vultures. Wintering species include Black-winged Kite and Marsh Harrier.

In winter, Ria Formosa is noteworthy for its huge wintering populations of waterbirds. Ducks can exceed 10,000, including Wigeon, Shoveler, Pintail, Teal and Pochard. Wildfowl is best looked for in pools and lagoons.

Waders include Avocet, Oystercatcher, Dunlin, Little Stint, Black-tailed Godwit, Spotted Redshank and many others common wintering species, but there are also a few Black-winged Stilts, Knots and Curlew Sandpipers. Scarce waders have included Temminck's Stint.

Other interesting waterbirds to be found in winter include good numbers of Greater Flamingos, Spoonbills, a few Glossy Ibises and some Caspian Terns, while the passerines include Water Pipit, Bluethroat, Penduline Tit and Reed Bunting. Slender-billed Gulls have been regularly recorded in small numbers during the winter months.

Caspian Tern

Accommodation

Algarve has the highest density of tourist accommodation in Portugal and therefore finding a place to stay should not be a problem, although in mid-summer it is well advisable to book in advance. In winter, it is usually easy to find accommodation at very low prices.

At Faro and Olhão there are numerous hotels and pensions to suit all pockets. The town of Loulé, lying about 17km to the north-west, also has several good hotels.

Strategy

This area is good for watching birds at all times of the year, except possibly in June and July because of the heat. Spring and autumn are the richest seasons in terms of variety of species, while winter is marked by huge concentrations of waterbirds. March and April are the best months to look for migrants and when the wind blows from the east for two days or longer, large numbers of migrants may arrive in eastern

Algarve presumably after crossing the sea or drifting westwards overnight from Spain.

Faro has a very convenient central location from where it is possible to explore either to the west or to the east. If coming by car, this town can easily be reached by following the A22 motorway.

a) West of Faro

In the western tip of Ria Formosa lies Quinta do Lago, a tourist complex with a golf course and a few small but important lagoons. A two-storey wooden hide has been built here. From this hide, Purple Swamphen, Little Bittern and a variety of waterfowl including Red-Crested Pochard can be observed at close quarters. Rarities here have included Red-knobbed Coot, White-headed Duck and Marbled Teal. The best access is from the village of Almancil. From there follow the signs to Quinta do Lago. On entering the tourist complex (clearly marked) follow the road straight through several roundabouts until you reach the sixth one – veer left and follow the Rua Douro for about 1km until a rough track sloping downwards appears on the right side. Take this track for 100m, park here and follow the marked 'discovery' trail to the left that passes along the marsh. This leads eventually to the actual lagoon at the edge of the golf course.

Ludo Farm is another excellent area close to Faro. Leave Faro westwards towards Loulé. Veer left after 3km following the signs to the airport and after about 1.5km take a road to the right. This road is in bad condition and after 600m deteriorates into a sandy track. Proceed along this track and you will soon see pines on the left and open fields on the right. A little further ahead, after passing through a gateway, there are

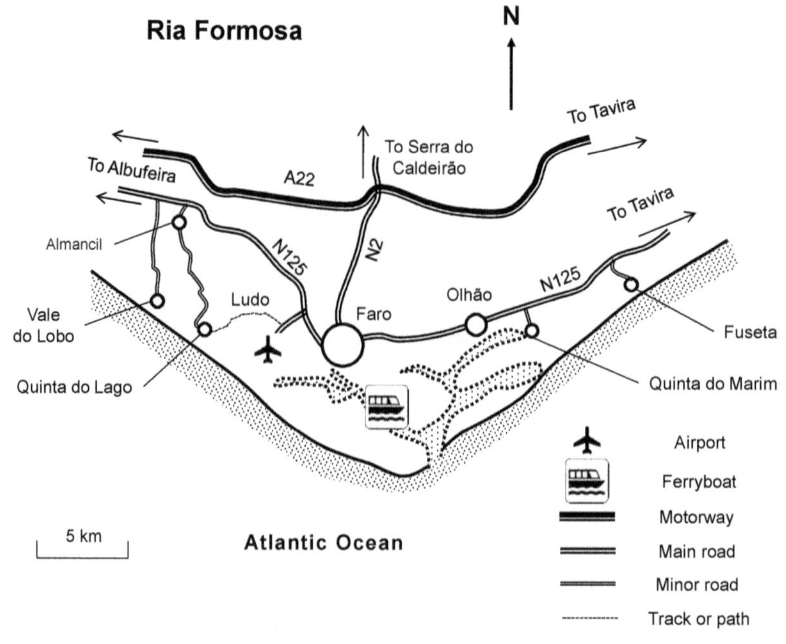

saltpans on the left, while on the right hand side there is a meadow, reedbeds and a pool, all of which can be good for wildfowl and for roosting egrets in the evening. The saltpans are good for wintering waterbirds, Osprey and Marsh Harrier, as well as for Bee-eater and breeding Little Tern.

The water front at Faro itself is a good and accessible place to birdwatch. The best place might be Parque Ribeirinho. At low tide there are usually a vast array of waders around and also herons, gulls and terns. This place can be reached following the signs to 'Parque Ribeirinho' from the roundabout at the south tip of Avenida Calouste Gulbenkian.

From Faro, one can take a boat trip on the channels of the Ria, where it is possible to watch several species of waterbirds. According to season, the list includes Spoonbill, Mediterranean Gull and Caspian Tern. On the islands there are good chances of Audouin's Gull in spring and summer. There are several companies operating boat tours and for most of them it is possible to get information at Faro docks.

b) East of Faro

The headquarters of the natural park (Parque Natural da Ria Formosa) is situated at Quinta do Marim, which is also a good site for watching birds. There is a 'Nature discovery trail', where it is possible to watch breeding birds like Purple Swamphen, Bee-eater and several passerines, as well as a few other species on passage. There are a few saltpans at the eastern end of the farm, which often hold parties of waders. Leave Faro on the eastern side, towards Olhão. Drive past the town, until a petrol station is reached 2.5km further east. Veer right just before this station, following the road to the headquarters of the Park. The railway line is crossed after 1km and there is a track to the left after 400m, where a gate marks the entrance of the Quinta do Marim. A small entrance fee is payable (€2.60 as of 2013).

Further to the east, the small fishing village of Fuseta is a good place to explore. From there it is possible to have easy access to the mud flats and also to a small complex of saltpans, which is worth inspecting for waders and gulls.

In the Olhão and Fuseta areas there are several companies offering birding tours both in the channels of the ria and offshore. Pelagic trips provide increased chances of seeing a few extra species, namely seabirds like skuas, petrels and even Sabine's Gull.

A few kilometres to the east lies the beautiful city of Tavira surrounded by saltpans complexes. It is one of the best sites in the Ria Formosa area and also one of the easiest to explore. There is a road that crosses one of these saltpan complexes and leads to the small port, from where the boat to Ilha de Tavira (Tavira island) departs. Without leaving the road one can see many waders at close quarters. Depending on the season, Audouin's Gulls in various plumages are usually around and there might even be the chance to see Slender-billed Gull. From the town follow the signs to 'Praia', 'Ilha de Tavira' and 'Praia 4 Águas'. This place can also be reached on foot from the town centre without too much effort. Close to Santa Luzia, a small place on the outskirts of Tavira, there is also an interesting saltpan complex.

Castro Marim

The Reserva Natural do Sapal de Castro Marim e Vila Real de Santo António (hereafter called Castro Marim) is a protected area, which covers an area of wetland, saltmarsh and saltpans, surrounding these two towns at the mouth of the Guadiana. The region has been occupied since Neolithic times; Phoenicians found the site to be fortified and Roman-Lusitanian and Mozarabic remains abound.

Although not very large, the estuary is very important for breeding, migrating and wintering waterbirds.

Location

Geographical coordinates: 37°13'06.00"N, 7°26'29.00"W

Looking at a map of Iberia, the best way to find Castro Marim is to follow the Guadiana to where it reaches the sea, at the south-eastern tip of Portugal. Full details of the area can be found on the following maps: 1:100,000: 50; 1:50,000: 50-D; 1:25,000: 591, 600

Description

The village of Castro Marim lies in the centre of the area and from its castle it is possible to get goods views over the surrounding wetlands.

To the south lies Vila Real de Santo António, once called Santo António de Arenilha, until it was inundated by a tsunami and rebuilt during the 'reign' of Marquês de Pombal in the 18th century.

This ecologically complex place has tidal changes and a fluctuating water table which result in a wealth of fish, molluscs and crustaceans, as well as a wealth of birdlife. The saltmarsh vegetation is rich in species whose life-cycle depends on intermittent winter rainfall.

Close to the river Guadiana and its two minor tributaries (locally called esteiros), the Lezíria and the Carrasqueira, the plant life is characteristic, forming vegetation bands parallel to the water, again much influenced by tidal ebb and flow. Large areas of the drier habitat called 'charcas' are bare of vegetation, saturated with salt and unusual in texture and composition.

Throughout the area there are several saltpan complexes, some of them still being used for the production of salt.

Birds

This site is not very far from the north-western coast of Africa and usually enjoys mild winters, so the area is extremely rich at that time of year. Regular wintering wildfowl and waders may include some Little Stints and Curlew Sandpipers, as well as hundreds of Black-winged Stilts, Avocets and Black-tailed Godwits. In cold winters, the area can hold parties of Black-necked Grebes and many dabbling ducks, as well as Pochard and Coot. Greater Flamingos are found all the year round, sometimes in large flocks. Spoonbill numbers fluctuate, and are highest in spring and autumn. Little Gulls, Caspian Terns and marsh terns occur in spring and autumn, the second named roosting on saltpan banks. Audouin's Gull is regular and hundreds of birds are sometimes around. Slender-billed Gull is rarer but the number of records has been increasing in the last few years.

The area is good for breeding birds. During the breeding season the saltpans hold Black-winged Stilt, Avocet and Kentish Plover, while Little Terns are common throughout the area. In terms of landbirds, larks are particularly noteworthy, as the place boasts at least five species: Thekla, Crested, Woodlark, Lesser Short-toed and Short-toed; the last named is a summer visitor and passage migrant. Additionally, Skylarks occur in winter. It is remarkable to watch Short-toed and Lesser Short-toed Larks in the same field of view and note just how different they are in plumage, behaviour and ecological strategy. The race of Lesser Short-toed Lark here is *apetzii*, a dark grey-brown subspecies, which makes identification rather straightforward. There is also a small population of Little Bustards and Stone-curlews.

Other breeding landbirds include Spectacled Warbler (which nests in small numbers in the vegetation along the saltpan edges, arriving in late February and departing in October), White Stork (breeding may begin in January), Red-necked Nightjar, Pallid Swift, Bee-eater, Black-eared Wheatear, Melodious Warbler and Southern Grey and Woodchat Shrikes.

Passage can also be exciting. Between March and May, Castro Marim holds thousands of migrant waders, as well as the last birds of the large winter populations; Spotted Redshanks, Ruffs, Little Stints and Curlew Sandpipers can be counted in their hundreds, some in superb summer plumage. Wood Sandpiper is regular and Temminck's Stint has been recorded. Garganey is regular in spring and autumn. Gull-billed Terns are scarce with odd birds in autumn mainly. Spring also brings migrating raptors with Black Kite, Montagu's Harrier, Short-toed and Booted Eagles. A short drive to the hills north of Castro Marim can produce wandering Griffon Vultures at times.

As to the landbirds, passage can bring Turtle Dove, Great Spotted Cuckoo (a very early spring migrant), hirundines, pipits, wheatears, Nightingale, warblers, Pied Flycatcher, and Ortolan Bunting. Though Iberian Yellow Wagtail is ubiquitous in summer, three other races – *flava, flavissima* and *thunbergi* – are regular on passage in spring and autumn though more difficult to identify at that season.

Greater Flamingo and Avocet

Accommodation

Accommodation should not be a problem in this area. It is possible to stay at Monte Gordo in reasonably priced hotels or at the camping site there. Vila Real de Santo António is also a good option. Another possibility is to stay at Tavira, about 20km to the west, which can be a convenient place from where to explore the Ria Formosa as well.

Strategy

Like most coastal wetlands in southern Portugal, this place is interesting at all times of the year, with the largest concentrations of birds outside the breeding season. Castro Marim village is within very short distance from the A22 and can be easily reached using this motorway.

a) Venta-Moinhos

This is the area east of the N122, north-east of Castro Marim and consists of saltpans, pasture and some open water pools. The best way to approach it is to follow the N122 up to Castro Marim village and proceed north on the road marked 'Beja'.

About 1.6km further north, close to the tiny village of Monte Francisco, take the dirt track on the right. This will lead you to the nature reserve headquarters after about 2km where the area of Venta-Moinhos begins. This place is good for Black-eared Wheatear in spring and Tawny Pipit in autumn. The track is an excellent route to take in the early morning during spring and autumn. It passes old abandoned saltpans and rough pasture. During spring and autumn passage the pools and saltpans can sometimes hold interesting waders like Temminck's Stint and overhead migrants can include Bee-eater, Pallid Swift, Short-toed Lark and Tawny Pipit.

Close to the nature reserve headquarters, the track branches within sight of the Guadiana and it is worth stopping here for a more protracted period. Just in front lie some flooded rushy pools, excellent for waders. In late April the site has held Temminck's Stint, Red-throated Pipit and the *thunbergii* race of Yellow Wagtail from Scandinavia. Just behind these pools lies dry salt pasture, which holds Little Bustard, Stone-curlew and Lesser Short-toed Lark throughout the year. At high tide, the extensive saltpans to the right of the track can fill with roosting waders and gulls. The saltpan banks hold breeding Spectacled Warbler in spring/summer. The track to the left leads to the base of the large bridge, which links Spain to Portugal. Here Montagu's Harrier can be seen in summer and passerine migrants are very common both in spring and autumn. The track to the right leads through pools and rough pasture where in spring Lesser Short-toed Lark can be watched closely, spinning its extraordinary mimetic song in a Serin-like song flight. Black-winged Stilt, Kentish Plover and Spectacled Warbler all breed here. Audouin's Gull, Gull-billed and Caspian Terns have all turned up in this area. Best strategy is to park the car at the parking lot of the reserve and explore on foot. At high tide this can be extremely profitable.

b) Castro Marim and surroundings

This is the area lying east of the N122 between Castro Marim and Vila Real. The bird life is similar to that of the previous place, but is less plentiful and is more difficult to observe. Fewer Lesser Short-toed Larks occur here, though in spring the site is filled with the crisp clipping notes of Short-toed Larks. Approached in the evening with the sun in the west, superb views can be obtained of Stone-curlews and displaying male Little Bustards. In summer, under some meteorological conditions, tens of thousands of swifts feed low over the ground on the swirling masses of aerial plankton glinting in the rays of the setting sun – the effect is indescribable. The area south and east from the village of Castro Marim is best reached by driving along the N122 towards Vila Real de Santo António. Left of the road several tracks run to the river bank (the bank of Carrasqueira). It is advisable to walk these tracks and not to enter the fish farms (easily identified behind high fences).

Just north-east of Castro Marim, along the N122, you can see abandoned saltpans. These are worth exploring taking the first mud track to the right after leaving the village. Walking is preferable here and highly advisable in wet weather. If the pans are wet, there can be good numbers of waders and even in winter Little Stints and Curlew Sandpipers are regular. In spring Spectacled Warblers breed here in the thick vegetation and juvenile Lesser Short-toed Larks (wandering from the main breeding site) can be found here in autumn.

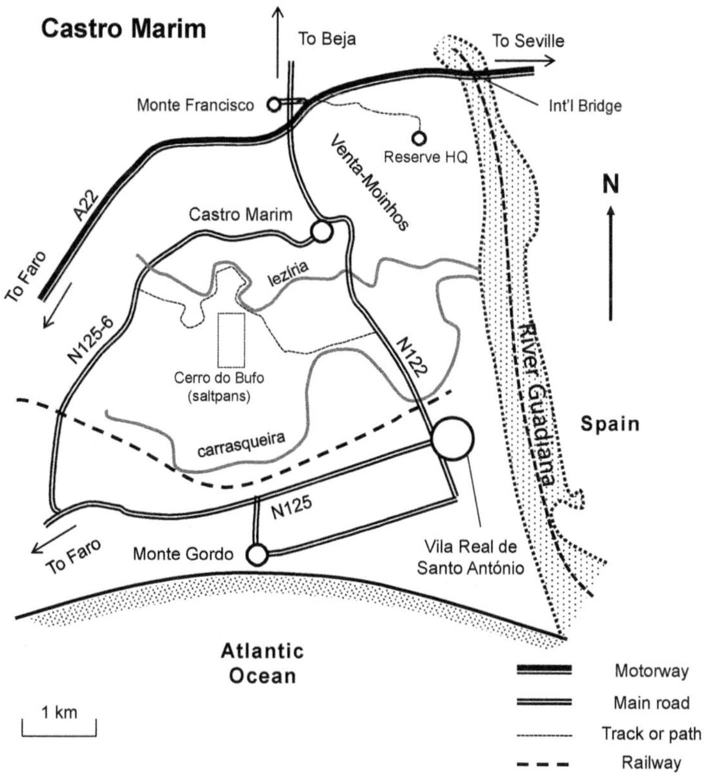

c) Cerro do Bufo

This sector lies between three roads: the N125 which forms the southern limit, the N125-6 which marks the western and northern limits and the N122 on the eastern side. It is a large area and worth careful, if time-consuming coverage. There are some access restrictions here and some places may be closed. It can be approached in two ways. One option, best taken early in the morning is to enter from the east. To do this, leave Vila Real de Santo António on the N122 northwards and turn left more or less halfway to Castro Marim down a short track to where a ruined farmhouse stands beside the esteiro da Carrasqueira.

The track can be rough and might be unpassable at certain times of the year. Therefore, it is preferable to park and walk westwards with the stream bank to your left and Olive and Carob groves to your right. White Storks breed here in exposed low trees and should not be disturbed. In spring the area can be filled with migrants and Black-eared Wheatear can usually be found near the ruined farmhouse. A host of interesting passerines pass through this site in autumn, many of them trans-Saharan migrants. It is worth glancing at the stream occasionally, as Red-knobbed Coot, Gull-billed Tern and even Long-tailed Duck have appeared here. The track continues for up to eight winding kilometres round the periphery of the saltpans, but the walk can be very rewarding with close-up views of Greater Flamingo, thousands of waders and Spectacled Warbler singing in spring. In winter Black-necked Grebes can be seen. Audouin's Gulls are usually around and Slender-billed Gulls can also turn up here. This track eventually leads to the salt storage area.

Another option is to access the salt storage area from the N125-6. Driving this road from Castro Marim towards Faro veer left just after crossing a metal bridge where Red-rumped Swallow sometimes breeds. Follow this track to the salt works gate, examining the pools on either side. Leaving your car here outside the gate, you have a choice of two parallel tracks. The inner one (just inside the salt works entrance) winds its way southwards along saltpan banks where literally thousands of waders and ducks can be seen from autumn through to spring. It is best traversed in the afternoon. Rarities here, among many interesting migrants, have included Wilson's Phalarope and Forster's Tern. The track ends at the bank of the esteiro da Carrasqueira near a white pump house. The channel here can be filled with wildfowl, though a telescope may be necessary to scan each saltpan bank carefully. The outer track leads to a pool that is often packed with waders and ducks when flooded. In summer it only holds breeding Kentish Plover and Spectacled Warbler.

AZORES

Lying in the middle of the North Atlantic, roughly half way between Europe and North America, the Azores archipelago consists of nine islands plus a few uninhabited islets. The exact date of its discovery is not known – officially, it was 'found' in 1431, although a few poorly documented versions claim that the archipelago had been known before this relatively late date.

The nine islands are grouped in clusters, usually called 'groups' (see map): the central group is the largest, with five islands; the eastern group has two, though one of them – São Miguel – is the largest and holds about half of the human population of the archipelago. Finally, the remote western group consists of just two small islands and has few inhabitants.

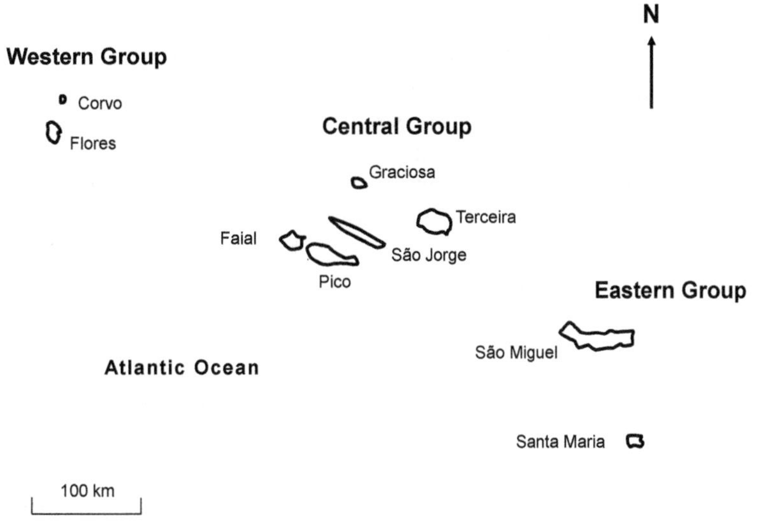

The Azores are of volcanic origin and the craters of old volcanos can be found throughout the archipelago – but most of them have now turned into lakes (known locally as 'lagoas'). The latest eruption that took place at Capelinhos in Faial began in September 1957 and lasted for 13 months. As a result, a new portion of land emerged.

Much of the original vegetation cover has long gone and now it can only be found in small patches mainly on São Miguel and in inaccessible slopes in some of the other islands. Large areas were cleared and the landscape is now dominated by farm pasture and plantations of exotic trees like Japanese Cedar (*Cryptomeria japonica*). Cattle grazing is one of the main activities.

As with most islands, the variety of resident birds on the Azores is relatively small, but there is an endemic species (Azores Bullfinch) and a number of endemic subspecies. In the case of Goldcrest, there are even three different ones! The archipelago is, however, extremely important for its breeding seabirds – about a dozen species breed, some of them in very large numbers.

Additionally, due to its relative proximity to North America, the Azores are probably the best place in the Western Palearctic to find Nearctic vagrants. This has led to a sharp increase in the number of visiting birders, who get to the Azores, especially in autumn, hoping to see some of those unusual American visitors.

Eastern Group

São Miguel

Most birders visiting the Azores arrive initially on São Miguel which is the main gateway to the archipelago. The administrative capital Ponta Delgada is located here.

Location

São Miguel lies about 1600km west of mainland Portugal. Full details of this island can be found on the following maps: 1:25,000 (series M889): 26, 27, 28, 29, 30, 31, 32, 33 and 34.

Description

São Miguel stretches for about 64km along an east-west axis and its width varies between 8 and 15km. It is a rugged island, especially at its eastern corner where Pico da Vara reaches an altitude of 1103m. Its volcanic origin is still well evident and there are several hot thermal springs (in Furnas these springs are used to cook typical dishes). The old volcanic craters are now extinct and lakes have formed in some of them. The climate is temperate wet and the higher parts are often covered with mist. The laurel vegetation, which originally covered most of its surface, is now almost gone. Only a few remains can be found in the eastern part around Pico da Vara and serra da Tronqueira. Over the centuries, many exotic plants were brought to the island and currently they form most of the vegetation cover.

Birds

São Miguel is probably best known for being the only place in the world where the rare Azores Bullfinch occurs. This bird can be found in the eastern part of the island (serra da Tronqueira). The list of resident landbirds is relatively short but includes most of the subspecies endemic to the archipelago. Woodpigeon is not uncommon and Grey Wagtail, Robin, Blackbird, Blackcap, Island Canary and Chaffinch (greener and more pallid than continental birds) are abundant. Goldcrest is represented by the endemic subspecies *azoricus*, known only from São Miguel. Long-eared Owl is resident and although not uncommon it is very hard to see. Quail, Snipe and Woodcock are not particularly scarce. Some exotic species have been introduced and Common Waxbill can now be found in several places while Ring-necked Parakeet occurs in some gardens at Ponta Delgada and may already be breeding. The only resident freshwater waterbird is Moorhen. As happens to be the case

Azores Bullfinch

throughout the islands, São Miguel also holds good breeding seabird populations, namely Cory's Shearwater and Macaronesian Shearwater.

Holarctic vagrants do occur with some regularity, especially in the post-breeding period. Waterbirds are the most likely species to be found. Several waders that are common in continental Europe are also quite regular here including Whimbrel, Turnstone, Sanderling and Greenshank. Some American vagrants too can be considered regular and among these there are species like Semipalmated Plover, White-rumped Sandpiper and Pectoral Sandpiper. Among ducks Teal is a regular winter visitor in small numbers and vagrants like Wigeon, American Wigeon and Ring-necked Duck are recorded with some regularity. Gulls can be common at places like Ponta Delgada harbour. The resident Yellow-legged Gull (race *atlantis*) usually dominates but other species also occur. Moreover, there is a long list of waterbirds that have been recorded on a few occasions only and the potential for surprises is very large.

Accommodation

São Miguel is a tourist resort and accommodation should not be a problem, although during the summer months it is advisable to book in advance. Most hotels are at Ponta Delgada but there are other possibilities in smaller towns and villages, namely rural houses that have been converted to small hotels.

Strategy

Almost all good birdwatching sites are reachable within one hour or less drive from Ponta Delgada. This means that this town can be a good basis to explore the entire island but a car is needed for this purpose.

a) Western sector

Ponta Delgada is not a very big town but it is the main one on the island. It has a wide harbour and sea front. This is worth exploring as it is one of the best spots in São Miguel for gulls. Some waders are also regular there and these include Dunlin, Sanderling and Knot. Among the rarities recorded here, the following should be mentioned: Common Scoter, Night-heron, Spotted Sandpiper and Bonaparte's Gull. This area can be easily explored on foot from the town centre. The gardens and parks of the town may prove interesting. Almost all passerines species breeding on the island can be seen there, along with Long-eared Owl and the exotic Ring-necked Parakeet.

To the west of Ponta Delgada lies one of the best-known places in São Miguel: the Lagoa das Sete Cidades, which is actually a complex of two lakes, one being green and the other blue. This is a major tourist spot but also a must for birders; it has a high potential to hold Holarctic vagrants. Being one of the most visited places on the island it should come as no surprise that the long list of rarities already seen there includes Nearctic duck and wader species. Additionally a few common waterbirds like Grey Heron, Little Egret, Coot, Moorhen and the odd migrating Greenshank are often around. The surrounding woods are good for seeing almost all the local landbirds and a few vagrant passerines have been recorded here. The introduced Common Waxbill can be also found on the lake shore. Access to the area is straightforward. From Ponta Delgada follow the R 1-1 westwards along the coast. Pass the airport and follow the signs to 'Lagoa das Sete Cidades'. The first recommended stop is at the bridge that separates the two lakes. From there it is possible to scan the lake searching for birds. It is also possible to explore the shores on foot.

Another good place not far from Ponta Delgada is Fajã de Cima. This is a very small lake lying within pastures and agricultural land some 2km north from the town. The Fajã de Cima has become in recent years a good spot to search for Nearctic waterbirds. Among many others, Blue-winged Teal, American Wigeon and Great Blue Heron have already been seen there. Access to the area can be made from the motorway running from the airport to Ponta Delgada by following the signs to 'Fajã de Cima'. After crossing the small village, follow the signs to 'Capelas' and 'Pico da Pedra' and then turn left to 'Capelas'. The place is reached after about 500m, at the first farm on the right hand side.

For those who like seabirds and seawatching, Mosteiros is one of the best places to visit. Located at the north-western tip of São Miguel this area is well known for its natural pools. However it is also one of the best seawatching places on the island. Cory's Shearwater is guaranteed while Great Skua and Gannet can also be seen. Apart from seabirds, the small pools that form at low tide are very good to look for waders, especially during migration times. Some Nearctic rarities have been found there. Among them are Semipalmated Plover, Least Sandpiper and Spotted Sandpiper.

b) Central sector

The central part of São Miguel is not as rugged as the western sector and has a few lakes that are well worth a visit. Lagoa do Fogo ('fire lake'), on the top of the ancient Fogo volcano, is one of them. This lake is often hidden in the mist but when visible it offers a spectacular sight. From a birding point of view, this can be considered as little explored. Even so, it appears to have good potential to receive vagrants and Black-necked Grebe, Spoonbill and Ring-billed Gull have already been seen there. In the surroundings it is possible to find a good selection of the local specialities. For instance, Grey Wagtail, Goldcrest, Island Canary or Chaffinch are all easy to find. This area is some 40 minutes drive from Ponta Delgada and it is quite easy to get there by following the signs on the motorway that runs from Ponta Delgada eastwards. It can be also reached from the town of Ribeira Grande on the north side of the island. Access on foot can be made from the first viewing point that is found if coming from the north or the third one if coming from the south.

Another interesting water body is the Lagoa das Furnas, which does not lie very far from Furnas village with its famous boiling water springs. It is another regular spot for vagrant Holarctic waterbirds. The best approach is from Caldeiras. These are the remains of old volcanic activity and can be easily reached from the R 1-1 close to Furnas (signposted). There is a car park there and exploration can be made on foot along the lake shore.

Close to Lagoa das Furnas lies yet another interesting area to explore known as Achada das Furnas. This site is composed of wet pastures with some small ponds and lakes which make it very good for a wide array of migratory species. Access can be made from the road connecting Furnas to Ribeira Grande via Pedras do Galego. About 6km past Furnas and about one km before the golf course veer left to a gravel track. A few hundred metres ahead there are two ruined houses. The surrounding fields and ponds are good for vagrant ducks, rails and waders. Proceeding on the track until the end you can follow the tarmac to the right for a few hundred metres and then veer right on the first track towards Pico do Gato. This area is best explored on foot.

On the southern shore of the island, a few kilometres east of Ponta Delgada, Caloura is an interesting and accessible place to look for terns and waders. Common and Roseate Terns are not uncommon during the breeding season and waders like Knot, Whimbrel or Grey Plover are regular on passage. Rarities for the Azores have included American Golden Plover and Redshank. Caloura can be reached at km 17, along the R 1-1.

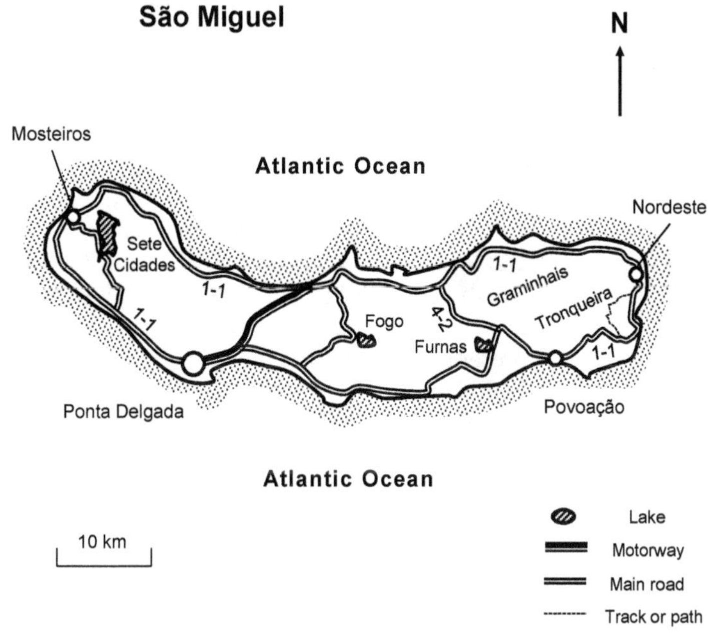

c) Eastern sector

The eastern sector of São Miguel is the place to look for the rare and threatened Azores Bullfinch locally known as the Priolo. This is a mountainous and rugged area where the largest remains of the original laurel forest survive. Until a few years ago access to this region was so difficult that it was sometimes called the tenth island of the archipelago. Nowadays, a new road has been built, allowing the journey from Ponta Delgada to Nordeste to be made in some 45 minutes.

As far as is known, the distribution range of the endemic Azores Bullfinch has never changed markedly and it has always been confined to the eastern part of São Miguel. The same cannot be said about its abundance. In fact, in the late 19th century the birds were reportedly so abundant that they were even considered as pests and a price was paid for their heads. Then, when the destruction of the habitat began, the population crashed and they became rare.

Fortunately, in the last few years a huge conservation project run by Sociedade Portuguesa para o Estudo das Aves (SPEA) has been in place and aimed at restoring the habitat. As a result there has been a significant population recovery. Seeing the Azores Bullfinch is now much more than just a possibility. As the birds tend to favour high altitude areas with natural forest, the best approach to exploring the region is to drive along the dust track known as 'rota da Tronqueira'. This connects the small Nordeste village to the R 1-1 a few kilometres east of Povoação. If coming from Nordeste the track can be taken from Pousada dos Clérigos (the hotel). If coming from Povoação on the R 1-1 eastwards, follow the signs to 'Rota da Tronqueira'. Besides Azores Bullfinch, almost all the other resident landbirds of the Azores can be seen along the road. A stop at the Tronqueira viewing point is highly recommended. Not only are the views breathtaking but also birding can be rewarding. The local race of Buzzard, Goldcrest and Island Canary can all be seen around here and in some winters there are records of Crossbills in this area. Another possibility to find the Azores Bullfinch is along the narrow tarmac road that connects the R 2-1 close to Furnas to the R 1-1 at Salga passing through the Salto do Cavalo viewing point.

SPEA runs a small but interesting interpretation centre about the Azores Bullfinch and the laurel forest; this centre is not far from Nordeste village at Pedreira (Centro Ambiental do Priolo – see the 'Useful Contacts and Websites' section at the end of the book). It is a pleasant place that is worth a visit and although the chances of seeing Azores Bullfinch here are remote there are however a few other resident Azorean passerines present. To get there follow the R-1-1 southward turning right at the Pedreira entrance towards 'Parque Florestal Cancela do Cinzeiro'.

Not far from Tronqueira lies a place known as Graminhais. This is a high altitude plateau some 900m above sea level which has an almost permanent fog cover with some open bog areas. Azores Bullfinch can also be found here and, besides, this is a good place to find breeding Snipe and Woodcock. Rarities like Wilson's Snipe and Naumann's Thrush have already been seen here. To prevent habitat destruction, some walkways have been built. If driving from Ponta Delgada to Nordeste, the best access to Graminhais is made via the small village of Achada. From there, one can follow a narrow tarmac road signposted 'Planalto dos Graminhais' that ends after some 9km.

Bear in mind that weather conditions at high altitudes can be adverse and can change very fast. Be prepared for rain, wind, cold and fog but also for some heat if the sky clears.

Santa Maria

Santa Maria was the first island of the Azores to be discovered and colonised. One of the reasons for this is its geographical location – it is the south-easternmost of the nine islands, lying about 100km to the south of São Miguel.

With its 97.4km^2 of land surface it is the smallest of the two islands that form the so-called eastern group of the Azores. Geologically, it is said to be the oldest of the archipelago and therefore it is the only one that does not show any signs of volcanic activity and also the only one which has some sedimentary formations. The coast has some pronounced bays and can be steep in some places but rocky stretches alternate with white sandy beaches (a rare feature in the Azores). At the south-western coast close to the main town of Vila do Porto there is a small islet called Ilhéu da Vila. Roughly it can be said that the western two thirds of Santa Maria are relatively flat and occupied by a low altitude plateau and the remaining one third at the eastern part is rugged, peaking at 590m at Pico Alto. The original vegetation has long gone and the cover is formed by a mix of woods (mainly Japanese Cedar plantations), pastures and agricultural land.

This island is one of the least visited by birders. This fact may be linked to its relative proximity to Continental Europe, which in theory reduces the possibility of receiving Nearctic vagrants. However, Santa Maria has much to offer from the ornithological point of view. It holds the largest breeding population of Kentish Plover within the Azores. It is also the only place on earth where one can find the endemic subspecies *sanctae-mariae* of Goldcrest. Besides, it has some important breeding seabird populations, namely at Ilhéu da Vila. These include Cory's and Macaronesian Shearwaters, Madeiran Storm Petrel, Bulwer's Petrel, Roseate Tern and the occasional Sooty Tern. Access to the islet is very difficult but it is in view from the main island.

Vagrants also occur. The best place to look for them is around the airport. The airstrip fields themselves can be good places but access is obviously not allowed. However views can be obtained from the perimeter and species like Hen Harrier, Kestrel or Peregrine Falcon (all of which are rarities for the Azores) have already been seen there. Outside this area there are a few temporary lakes that are worth checking for waterbirds. Killdeer has been recorded here during the breeding season in recent years. The best strategy is to walk along the fence and explore the surrounding areas. Other good places to look for vagrants are Baía dos Anjos on the north-west part of the island and the mouth of the ribeira de São Francisco at the south coast, close to Vila do Porto.

Santa Maria can be reached by air from São Miguel – there is at least one daily connection.

Central Group

Terceira

It is likely that Terceira owes is name to the fact that it was the third island in the Azores to be discovered and colonized (Terceira is the Portuguese word for third). But although it was not the first to be reached by the ancient navigators, it was probably the first one to be 'discovered' by modern birdwatchers keen to add Nearctic birds to their Western Palearctic lists.

Location Terceira is the westernmost island in the central group. Full details of this island can be found on the following maps: 1:25,000 (series M889): 22, 23, 24 and 25.

Description

With a land surface of 402.2km^2 this is the third largest island of the archipelago. It stretches for a maximum of 29km in length and its maximum width is 18km. The landscape is diverse including plains like Achada and mountains like Santa Bárbara that reaches 1021m, forming the highest point on Terceira. There are also a few lakes like Ginjal and Junco and one of the few true marshy wetlands on the Azores – the Paul da Praia da Vitória.

The western part has good vegetation cover dominated by exotic Japanese Cedar plantations. Besides, as in most other islands in the archipelago, large areas are occupied with pastures and agricultural land.

The island holds a population of more than 55,000 people (nearly one quarter of the Azorean population). The main towns are Angra do Heroísmo (its old historic centre being a World Heritage Site, well worth a visit) and Praia da Vitória.

Birds

Terceira is not much different from the other islands in terms of the landbird community and almost all the Azorean species can be found. Among the most interesting are Quail, Long-eared Owl, Woodcock and Island Canary. Additionally, it is one of the few places in the Azores where Kentish Plovers breed and it also holds an important population of Moorhen – a rare species in the Azores. As happens elsewhere in the archipelago, Cory's Shearwater is a common breeding species. Furthermore, there is a breeding population of Common and Roseate Terns.

This island is probably best known as a hot spot for vagrants particularly in the eastern part, around Praia da Vitória, where most of the suitable habitat can be found. Nearctic ducks seen included American Wigeon, Blue-winged Teal and Ring-necked Duck, just to name a few. Nearctic waders sights are not uncommon and some species like Semipalmated Plover, Pectoral, Semipalmated, White-rumped and Spotted Sandpipers, Lesser Yellowlegs and Wilson's Snipe can even be considered regular. The list of gulls is also impressive and Ring-billed, Bonaparte's, Mediterranean, American Herring, Common, Glaucous and Iceland Gulls have all been seen. Apart from these, the list of vagrants is long and although is not possible to mention them all here it is worth mentioning a few like Long-tailed Duck, Black-necked Grebe, Great White Egret, Glossy Ibis and Red-footed Falcon.

Accommodation

Accommodation should not be a problem on Terceira and there are many places to stay. However, to be on the safe side, it is best to book in advance, especially during summer months.

Strategy

Most of the key birdwatching sites lie on the eastern half of the island, where the two larger towns are also located. The best strategy is therefore to choose one as a base. A car is essential to move between the different places although that can also be achieved by taxi or even, in some cases, by using the public transport network.

a) Praia da Vitória and its surrounds

Praia da Vitória is the second largest city on Terceira and the one lying closest to Lajes airport (just 4km). It is a good starting point from which to explore the island. The marina, the main beach and the bay are good places to find gulls. Among the flocks of the common resident Yellow-legged Gulls (race *atlantis*) there are sometimes other gulls to be found. The list of species already seen here is pretty impressive and includes not only Nearctic species but also Western Palearctic ones. The best time to explore the area is towards evening when birds come to roost.

About 2.5km from Praia da Vitória, at Cabo da Praia, lies the fishing harbour. This is another good place to search for gulls and terns, among others. To get there, just follow the signs to Cabo da Praia. The harbour is not signposted but it is quite obvious.

Cabo da Praia is probably the most famous birdwatching site in the Azores. It is not a very pretty place in terms of landscape, as a set of fuel deposits have been built nearby and the area itself is in fact an old quarry by the sea, partly influenced by the tides. However, this location is particularly interesting for the birdwatcher. At high tide, ephemeral pools appear and these attract waterbirds. The list of vagrants seen here is long and includes many Nearctic waders. Besides rarities, this is one of the few places in the Azores with a breeding population of Kentish Plover. Cabo da Praia is only 3km away from Praia da Vitória. Access to the area is easy: at the end of the motorway connecting Angra do Heroísmo to Praia da Vitória one must follow the signs to Cabo da Praia proceeding then to 'Zona Industrial' (industrial zone).

Within the urban perimeter of Praia da Vitória there is a marsh known as Paul da Praia da Vitória. Despite heavy human pressure, the marsh has managed to survive and it was recently 'restored' by local authorities; a path and two small hides have been created and can now be used by visitors. Moorhen is not uncommon here and has become the symbol of this area. Furthermore, other waterbird species occur regularly, including Teal, Grey Heron, Little Egret and Coot. This is also a good place to find Nearctic ducks and American Coot and Belted Kingfisher have also been recorded.

Ponta das Contendas is a coastal site halfway between Angra do Heroísmo and Praia da Vitória. This location holds important breeding colonies of Common and Roseate Terns and has become a Special Protection Area (SPA). Apart from the seabirds it is also a nice place to find a good variety of local resident species. The best access is via the coastal road, turning to the Contendas lighthouse. About 1km after the lighthouse there is an iron gate on the right side. After passing through it and walking for a few hundred metres it is possible to reach a suitable watching point.

b) Angra do Heroísmo and surroundings

Start by exploring the bay of Angra do Heroísmo itself with its marina and walkways where locals like to stroll. Although it can seem uninteresting for birdwatching purposes, it is worth a look. Migrating waders like Turnstone, Sanderling and Whimbrel are often around, the same applying to Little Egret and Grey Heron. The place became famous when the first Yellow-crowned Night-heron for the Western Palearctic was found there in late 2009. Other vagrants have been recorded here, namely Double-crested Cormorant, Osprey, Peregrine Falcon and Spotted Sandpiper.

About 11km to the north of Angra do Heroísmo lies Lagoa do Ginjal. This lake is surrounded by high peaks and somewhat 'hidden'. It has almost no vegetation along its banks. This is one of the best birdwatching spots on the island. Moorhen breeds there and many other Azorean breeding species can be seen around. The list of vagrants is extensive. Access to the area is not straightforward. It can be made from the motorway that connects Angra do Heroísmo to Praia da Vitória, leaving at the exit to 'Biscoitos', 'Cabrito' and 'Zona Industrial'. More or less in the middle of the industrial area, turn right and then left and proceed until you find a track to the right that will lead to the lake.

Not far from Lagoa do Ginjal lies Lagoa do Junco, half way between Angra do Heroísmo and Praia da Vitória. Until recently this was an underwatched area but the number and quality of records made there in recent years have turned it into one of the great spots of the island for vagrants. Although the place has a high potential bird-wise, it suffers from disturbance caused by agricultural activities and especially hunting. In the surroundings there are vast pastures and in autumn and winter there are many small pools.

The best access to this site is from the motorway connecting Angra do Heroísmo to Praia da Vitória. More or less 15km from each, follow the signs to 'Estaleiros', 'Porto Judeu' and 'Serra do Cume'. After turning there it is possible to follow either the right or left hand roads as both cross areas with small pools before reaching the lake itself.

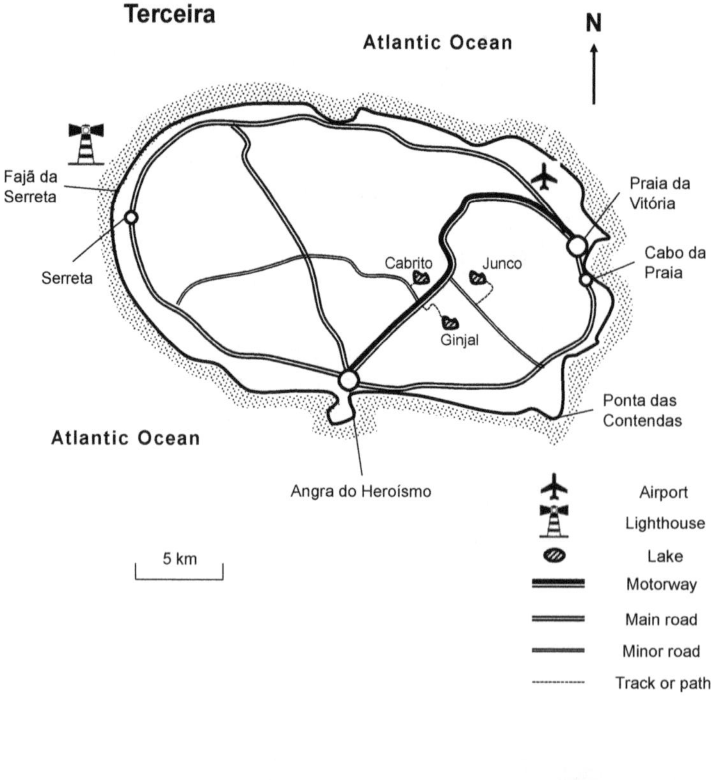

Cabrito reservoir is an artificial lake and it does not have a very attractive setting. However, during autumn and winter it deserves a visit, as it can attract Nearctic ducks and waders. Species already seen include Ring-necked Duck, American Wigeon and Lesser Yellowlegs. The access is also from the motorway. About 8km after Angra do Heroísmo leave it heading towards 'Cabrito'. The reservoir is 1km beyond.

c) Western side

Compared with the eastern half of the island, the western side is not so exciting from a birding point of view, mainly because there are not many wetlands. One of the few interesting places is Fajã da Serreta. This is an agricultural area with some patches of exotic vegetation and also small patches of native coastal vegetation. It is a good place to look for resident Azorean bird species. Landbirds like Quail, Buzzard, Woodpigeon, Grey Wagtail, Island Canary, Goldfinch or Greenfinch, among others, are not uncommon. Cory's Shearwater is a common breeder in the cliffs. The area has good potential in terms of American passerines but it is underwatched. There is one record of American Ovenbird from here, underscoring its undoubted potential.

This place is located some 20km west of Angra do Heroísmo. Take the regional road to Serreta and at the end of the village turn left towards the lighthouse and Ponta da Serreta headland. About 600m ahead there is a track to the right that will lead to Fajã da Serreta in 800m. This track can be in bad condition, so it is not advisable to drive along it.

Graciosa

Graciosa is the northernmost island of the so-called central group of the archipelago. It is not very large, is oval-shaped and has a total land surface of about 62km^2 with a maximum length of 12.5km and width of 8.5km. Despite being of volcanic origin, it is the least rugged of the Azorean islands and the maximum altitude is reached at 402m at the eastern side of the Maciço da Caldeira, located at the south-eastern tip. This general low altitude is responsible for the mild oceanic climate and for the lowest rainfall of the central group. Most of the nearly 4400 habitants live at Santa Cruz da Graciosa, the main settlement on the island.

The landscape of Graciosa is a mix of pastures and agricultural fields. The coastline is generally low, rising only at the south-east stretch of serra Branca. There are two small islets not far from the coast (Ilhéu da Praia and Ilhéu de Baixo) both holding good numbers of breeding seabirds.

This is undoubtedly the least visited island for birders. However, it is probably the most interesting one for seabird watching. It is likely to be the best place in the world to find the rare and recently split Monteiro's Storm Petrel, an endemic for the Azores (which is, however, not recognized by some authorities) that breeds during the summer season. Madeiran Storm Petrel also occurs but it breeds in winter. Furthermore, Cory's Shearwater is a widespread breeder and Macaronesian Shearwater and Roseate Tern breed on the islets. Among the landbirds it is possible to find a few resident species like Quail, Buzzard, Woodpigeon, Grey Wagtail, Blackcap, Starling and Island Canary. Some waders turn up along the coast and Turnstone, Whimbrel, Grey Plover, Knot and Sanderling, to name a few, are more or less regular. Although as stated before this island is probably underwatched, there are records of rarities like Great White Egret, Spotted Sandpiper and Purple Sandpiper. Snow Bunting has been recorded in winter at serra Branca.

The easiest way to get to this island is by air – there are daily flights from Terceira. Exploration of Graciosa is fairly easy. From the airport to the centre of Santa Cruz (at the north-western tip) is a 5 minute drive. Once there it is possible to explore on foot starting with the so-called 'pauis' (two water reservoirs inside the village itself) then proceeding to Calheta up to Santa Catarina along the coast. This walk can allow for the observation of some waders, gulls and, occasionally, the odd egret. The peak of serra Branca lies about 5km further south. Apart from these, there are a few marked paths that can be used for exploration (on foot) of the inland section of the island.

São Jorge

Looking like a cigar on the map, São Jorge has a distinct shape, being long and narrow. Although probably not the best island of the archipelago in terms of breeding birds or rarities, a visit might prove rewarding as this place is very beautiful in terms of landscape.

Despite its narrowness, São Jorge is quite high, peaking at just over 1000m. The whole coastline is very rugged and along it there are several accumulations of volcanic boulders, called Fajãs. On the higher parts of the island most of the vegetation has been cleared for agriculture, but on the slopes there are still good patches of natural forest.

The variety of resident landbirds is similar to the other islands of the central group. The Azorean species of Buzzard is quite common and so is the Island Canary. Other birds that can easily be found include the local races of Goldcrest (*inermis*), Chaffinch and Grey Wagtail.

There are large numbers of breeding Cory's Shearwaters, with the largest colonies on the islets at the eastern and western tips. Other breeding seabirds include Macaronesian Shearwater, Common and Roseate Terns, but these birds are less numerous here than on other islands, such as Flores, Graciosa or Santa Maria.

During migration times, unexpected vagrants can turn up. Unlike most other islands, São Jorge has no lakes in the middle. However, the two Fajãs on the northern side (Fajã dos Cubres and Fajã da Caldeira do Santo Cristo) have a small lake each and are the best place to look for migrant and vagrant waterbirds. Rare American vagrants that have been recorded at Fajã dos Cubres include Blue-winged Teal, Surf Scoter, American Bittern, Little Blue Heron and Common Nighthawk among others.

São Jorge has a small airport which is served by daily flights from Terceira or São Miguel. It is also possible to get there by boat, from Pico or Faial.

Like elsewhere on the Azores, a hired car is essential to travel around. Fajã dos Cubres lies on the northern shore, close to the village of Norte Pequeno, but the other Fajã with a pond can only be reached on foot (40 min walk from the former). Other interesting places to visit are the eastern and western tips of the island, which can prove rewarding for

Cory's Shearwater

Pico

Pico gets its name from the pointed mountain that rises from the sea and peaks at 2351m, forming the highest point anywhere in Portugal. This island is the second largest in the Azores.

Location Pico lies in the middle of the so-called central group, between Terceira, Faial and São Jorge. Full details of this island can be found on the following maps: 1:25,000 (series M889): 7, 8, 9, 10, 11, 12 and 13.

Description Despite its large size (it covers an area of 447km^2) Pico has a rather small human population (about 14,000 inhabitants in 2011 and declining). There are three main settlements all lying on the coast: Madalena, facing west, is the entry point for those arriving by boat; São Roque do Pico, on the northern part of the island, is an important fishing harbour; and Lajes do Pico, which stands on the southern shore. Any other settlements lie mainly along the coast, the inner part being very sparsely populated. There are some patches of woodland, together with agricultural fields (including the vineyards which have been recognized as a World Heritage Site) and a few freshwater lakes, of which Lagoa do Capitão and Lagoa do Caiado are the most important.

Although the island has an airport, there are not many flights and so for most visitors the most convenient way to reach Pico is by boat from Faial – a 20 minute crossing which is usually interesting for seawatching.

As seen from Faial, the mountain looks very symmetrical, but as one proceeds eastwards the symmetry disappears.

Birds Pico's assortment of breeding landbirds is not much different from that of the other islands in the central group – there are good numbers of Buzzards, Woodpigeons, Goldcrest and Island Canary, alongside the widespread Robin, Blackbird, Blackcap, Starling and Chaffinch. In the higher areas in the central part of the island it is possible to find breeding

Woodcock and Snipe (these species are possibly easier to find here than on other islands) and there have been a few records of Wilson's Snipe.

The variety of seabirds is somewhat smaller compared with Graciosa or Santa Maria, for instance, but as elsewhere there are large numbers of Cory's Shearwaters, along with a few colonies of Macaronesian Shearwater and Roseate Terns. Additionally Pico is well-known for seawatching (namely whale-watching) and has proved interesting for seabirds during pelagic trips. Inter-island crossings (namely to and from Faial or São Jorge) also provide seawatching opportunities.

Pico has its share in terms of rarities, which can turn up anywhere and it probably has remained under-explored, as many observers skip visiting this island entirely. Several American waders have been recorded and some species, like White-rumped and Spotted Sandpipers have been seen regularly, but very rare species have turned up on many occasions. Other Nearctic birds that have been recorded on Pico include American Wigeon, Blue-winged Teal and Ring-necked Duck, along with Little Blue Heron, American Bittern and American Coot.

Accommodation

In terms of accommodation Pico is not very well developed and this can mean that it is hard or impossible to find a place to stay in the high season. It is thus highly advisable to book in advance, especially during the summer. Another possibility is to search locally for a private room.

Strategy

The best way to explore Pico (and certainly the most convenient for those arriving by boat from Faial) is to rent a car and make a circuit from Madalena, passing through Lajes, the lakes and finally São Roque. If arriving by air, it should not be difficult to drive the same circuit by hired car.

Start by exploring the harbour of Madalena. Several rarities have been recorded here, including many waders and the Nearctic Little Blue Heron.

Leave Madalena on the road N1 that follows the south coast eastwards. After about 40km one arrives at Lajes do Pico. This village takes its name from the lava fields that have formed on the sea shore. This is one of the few locations on the island where the coast is not rugged and which provides a good feeding place for wading birds and so it is one of the better places to look for Nearctic waders.

In the central part of Pico, east of the mountain, there are several small lakes which are worth checking especially for vagrant waterfowl. The largest and most accessible ones are Lagoa do Caiado and Lagoa do Capitão. Species recorded there included vagrant waterfowl and waders. Breeding birds in this area include Snipe and Woodcock. For Lagoa do Caiado, leave Lajes do Pico northwards on the N2 and turn right after about 7km; proceed for another 3km. For Lagoa do Capitão, proceed on the N2 for another km and veer left taking the N3 towards Madalena. The lake will appear on the right side after about 2km. It should be noted that weather conditions can be bad here – because of altitude, fog is often present; do not forget to gear up accordingly.

Several other lakes exist further east, such as Lagoa do Paul, Lagoa do Peixinho and Lagoa Rosada, where the avifauna is similar to that of the above-mentioned lakes, but a detailed map is essential to get there.

The northern coast is also worth a visit. The main settlement of São Roque, a former whale centre, has a small harbour which is worth checking for any gulls or waders.

Pico

Faial

Despite being one of the better known islands in the Azores – especially because of its popular harbour paintings, the famous Peter Café Sport and the eruption of the Capelinhos volcano half a century ago – Faial is not a prime site for the birdwatcher, largely because there is not much suitable habitat to attract 'interesting' species. Given its geographical location at the western extreme of the central group, this island is essentially a convenient stopover for those travelling by air to Flores or by boat to Pico.

Faial is rather circular and somewhat intermediate in terms of steepness – it is not as rugged as Pico or São Jorge, but not as flat as Graciosa or Santa Maria. In the middle of the island lies an ancient crater.

Horta is the main town. Although quite small it is a pleasant place. Elsewhere the island is mainly green, the landscape being dominated by pastures and agricultural fields, with a few patches of woodland here and there. The western tip (Capelinhos) has been formed recently and lacks vegetation of any type. There are hardly any water bodies, except for a small lake near the village of Pedro Miguel and a lake at the bottom of the crater.

In terms of landbirds, the variety is similar to that of the other islands in the central group. Buzzard, Woodpigeon, Grey Wagtail, Goldcrest, Chaffinch and Island Canary are all fairly common and easy to locate, while Long-eared Owl is also present but is, as usual, more elusive.

The variety of seabirds is less striking that elsewhere in the Azores; nevertheless Faial holds several thousand pairs of Cory's Shearwater (which can be seen in good numbers from the ferry to Pico) and a few pairs of Roseate Tern, although this species does not breed every year. The Capelinhos area is good for both species.

Several rarities have been recorded here and Faial definitely has potential for vagrants and has been underwatched. For those aiming at finding some oddities, the best places are: the harbour of Horta, where Red-necked Grebe has been recorded; the bay of Porto Pim (about 1km south-west of Horta), which is one of the best sites for waders; the lake near Pedro Miguel, which is probably the best place for waterfowl and the rim of the crater, where unexpected birds have turned up, such as Dotterel and Snow Bunting.

Western Group

Flores

Flores is the largest of the two small islands that form the so-called western group of the Azores, which lies more than 200km from the central group and is actually located midway between Europe and North America. The island has an area of 143km^2 and despite its small size it is quite high (almost 900m at the top). Like most other islands in the Azores, Flores is quite rugged and the coastline is not very attractive to feeding waterbirds. However, the central part is occupied by a plateau where there are several lakes. The natural vegetation has long gone replaced by pastures and plantations of the exotic Japanese Cedar.

Because of its remoteness, Flores was not much visited by birdwatchers until the turn of the millennium. However, in recent years, there has been an increase in interest from European birders, who go there mainly in search of American vagrants.

The number of resident landbirds is smaller than in the rest of the Azores and the larger species, such as Buzzard and Long-eared Owl are absent, the same applying to Robin and to the introduced Greenfinch. Among the few species occurring, the most interesting ones are Goldcrest (represented here by the endemic subspecies *inermis*), Island Canary and Chaffinch (of the Azorean race *moreletti*), all of which are common and widespread. A small population of Black Ducks has been recorded regularly in the lakes and breeding has been proved on two occasions, but the genetic purity of these birds has not been ascertained.

The island is also very important for its seabird colonies: Cory's Shearwaters are numerous and there are colonies of Manx and Macaronesian Shearwater, as well as Common and Roseate Terns.

Autumn turns out to be the most interesting season because a large number of migrant birds turn up here and many of them are of American origin. Ducks and waders are among the groups with the largest number of records.

Getting to Flores is not complicated as there are daily flights from either Faial, Terceira or São Miguel (but it should be noted that the number of flights drops considerably outside the summer season). The airport lies on the eastern part, near Santa Cruz das Flores. Once on the island, a hired car is the most convenient way to explore (but in summer reservations should be made well in advance, as the number of cars for hire is limited). All the existing seven lakes are easily reached by car and they are all worth inspecting, for it is there that vagrant waterbirds are most likely to turn up. Of these, the shallow Lagoa Rasa seems to be more favourable for waterfowl and waders. It is located about 5km north-west of Lajes das Flores, not far from the road that links Lajes to Fajã Grande. It is also advisable to take a look at Fajã Grande, on the western coast, as several Nearctic waders have been recorded there.

Corvo

The smallest island in the Azores, with only about 430 inhabitants, Corvo remained relatively unknown birdwise for a long time but regular visits in the last few years have proved that this place has a very high potential for American vagrants. Corvo's reputation is now well established among Western Palearctic birders.

As with Flores, there are few resident birds – mainly passerines, most of which belong to endemic subspecies: Grey Wagtail, Blackbird, Blackcap, Chaffinch and of course the Island Canary. However, Goldcrest is absent here.

There are breeding populations of Cory's and Macaronesian Shearwaters, as well as Common and Roseate Terns. There are also several pairs of Manx Shearwater, which breed only in the western group.

However, it is the possibility of finding American vagrants that keeps attracting birdwatchers from Europe to this remote island, especially in autumn. Because of its small size and of its relative proximity to North America, Corvo has turned up a remarkable list of extra-limitals.

Getting to Corvo can be difficult. There are flights from Terceira, Faial and Flores, however these flights are not daily; furthermore, when the weather is very bad there may be no flights at all. It is also possible to try to hire a boat trip from Flores, but these will only be made when the sea is not too rough. Additionally, accommodation facilities are very limited and may be fully booked at the height of the migration season (mainly in early autumn). Given its small size, Corvo is not difficult to explore and that can be done on foot. Small vagrants can turn up everywhere, and seem to be easier to find on Corvo because of its smaller size. The only human settlement is Vila Nova do Corvo, which has about 400 inhabitants and lies on the southern tip of the island, where the airstrip is also located. In the northern half there is a large crater, on the bottom of which there is a lake. The place is attractive for vagrants and many oddities have turned up here. It can be reached on foot only (but it is possible to drive up to the rim of the crater on the eastern side).

It should be noted that weather conditions can be quite adverse in Corvo – storms are frequent and sometimes fog covers the entire crater. It is thus advisable to gear up for bad weather.

MADEIRA

The Madeira archipelago was discovered about 600 years ago, when Portugal led Europe in navigational skills and dangerous journeys.

Often called the 'Pearl of the Atlantic', the autonomous territory of Madeira actually consists of two main islands (Madeira and Porto Santo) and two clusters of smaller islands, Ilhas Desertas and Ilhas Selvagens, respectively (see map).

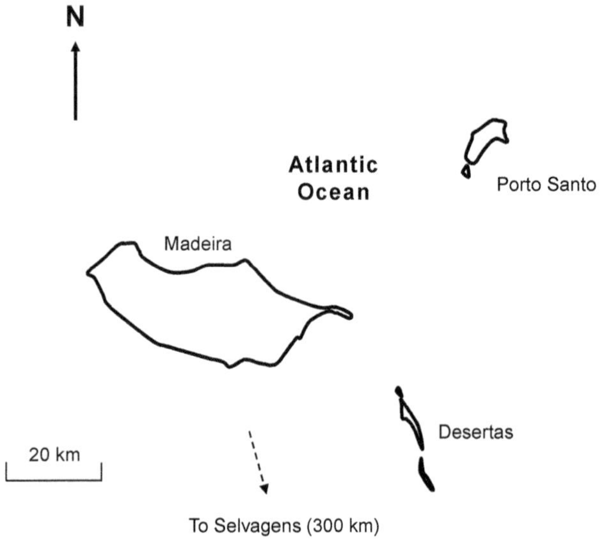

As elsewhere in the Macaronesian region, the archipelago has a volcanic origin. However unlike in the Azores or the Canaries, there has been no volcanic activity for many centuries.

Vegetation cover varies among the islands. Madeira (the Portuguese name for wood) is mostly green with a large area of laurel forest and many exotic trees while the other islands are barren and largely treeless.

The territory is famed for its seabirds, many of which it shares with other Macaronesian archipelagos, like the Canary Islands and the Azores.

Madeira

The main island of Madeira archipelago is a major tourist destination attracting people from all over the world. It is a pleasant place with mild climate and dramatic landscapes. Birdwise, Madeira is probably best known for its seabirds; however the endemic landbirds, both species and subspecies, are of considerable interest as well.

Location Madeira lies in the Atlantic Ocean some 850km south-west of mainland Portugal and 660km from the African west coast. Full details of this island can be found on the following maps: 1:25,000 (series P821): 1, 2, 3, 4, 5, 6, 7, 8 and 9.

Description

Madeira is one of the biggest islands of the biogeographic region known as Macaronesia. It occupies about 740km^2, it stretches for 55km (west to east) and its maximum width is 24km. The varied landscape of Madeira reflects its pyroclastic volcanic origin, a landscape of dark rocky vistas, peppered with extinct volcanoes, precipitous ravines and misty heights. The land slopes upwards markedly from a sheltered southern shore to an east-west backbone of rugged peaks and staggering cliffs. Pico Ruivo rising up to 1862m is the highest mountain in the region and the third highest in Portugal.

The climate is subtropically humid and is dominated by the north-east trade winds from Africa. On Madeira these climatic conditions are much altered by rapid changes in relief. Furthermore, there is usually a contrast in the weather between the northern and the southern side of the island: the former is wet and misty, whereas in the latter area the sun shines more often, sometimes registering very high UBV values. Together with human intervention, this produces a south-north zonation from southern shore banana plantations, through characteristic terraced small-holdings to orographically generated subtropical laurel forest, constantly swathed in rolling mist.

The Parque Natural de Madeira (a protected area) covers about two-thirds of Madeira. Its main concern is preserving the remaining 10,000ha of indigenous laurel forest or 'laurisilva' with its endemic flora and fauna. It is worth remembering that the laurel forest occupies an eighth of Madeira, forming the largest area of this type of habitat anywhere in the world.

Hydrographical studies in Madeira have identified over 200 waterways flowing into over 100 river basins. These have been canalised to irrigate the terraces that are so characteristic of the Madeiran landscape. The channels are cut-away water courses only a few metres wide and alongside these 'levadas', some of which tunnel through mountains, are narrow pathways.

Madeira is a densely populated island, especially on the southern half where the capital city Funchal is located.

Birds

For birdwatchers brought up in Continental Europe, the seabirds of the Macaronesian archipelagos have a quality almost beyond words. Their very names carry a nomenclatural charisma, often being called after the remote islands on which they breed or after early naturalists who explored this area. Madeira is the only place in the world where the threatened Zino's Petrel breeds. Much more common are Cory's Shearwater and Manx Shearwater. Apart from these, the surrounding waters boast species like Bulwer's Petrel, Macaronesian Shearwater, Madeiran Storm Petrel and Fea's Petrel (which some authors consider a separate species, Desertas Petrel).

The list of landbirds is not extensive but includes some special and specialized species. There are two Madeiran endemic species both closely associated with the island's laurel forest: Madeira Laurel Pigeon and Madeira Kinglet. The latter has been recently recognised as a full species. Furthermore, it is possible to find three other endemics – Macaronesian ones including Plain Swift, Berthelot's Pipit and Island Canary. The list of resident species includes other interesting species, like Spectacled Warbler and Spanish and Rock Sparrows, along with several more common birds, most of which are represented by local subspecies that may be taxonomically split in the future. Pallid Swift is a summer breeder. In the beginning of the 21st century Siskins were found breeding in coniferous plantations in the higher parts of Madeira.

Although not located on a major landbird migratory route, stragglers from mainland Europe sometimes turn up during the migratory periods. That is the case of a few waders that can be seen in very small numbers along the coast.

There are at least two introduced exotic species, both uncommon. One is the Common Waxbill which apparently is already established on Madeira. The other is the Ring-necked Parakeet, regularly seen, especially in Funchal's gardens.

Long-toed Pigeon

Accommodation

Madeira is a tourist honeypot and there is a host of hotels and pensions. Accommodation should not be a problem except maybe during the summer months or at Christmas/New Year. Advance booking through a travel agency is advisable.

Strategy

The best time to visit Madeira is between April and January but the endemic species like Madeira Laurel Pigeon, Madeira Kinglet, or the local races of Blackcap, Spectacled Warbler and Chaffinch can be found all year round and most of the exciting seabirds can be seen between June and September. Migrant landbirds like some birds of prey, swallows, pipits and wagtails are best seen in March, April and September but like all islands, Madeira may turn up all kinds of oddities. Its position off the African coast only adds to the excitement of finding the unexpected.

It is possible to use the public transport system to explore some of the sites but before doing that a timetable (horário) should be carefully studied. Taxis are widely available but it is advisable to calculate the costs and write them down before setting out. Car-hire is widely available booking being possible at the airport or arranged in Funchal or at the major hotels. Hiking is one of the major attractions of this island. The 'levadas' form a network of strange and mysterious byways that can be explored for many kilometres (rights of way).

a) Funchal

The capital of the island is a busy town where most of the visitors to Madeira stay. Even so, it offers some birding opportunities. Walking by the sea along Avenida do Mar can be rewarding. Gulls and Terns roost here, including *atlantis* Yellow-legged Gulls, Common Terns and sometimes Roseate Terns. The walk from Avenida do Mar to the outer wall passes over the ribeira de São João where Nearctic waders and gulls have been seen. Other rarities have been seen here and included species like Night-heron, Spoonbill, Glaucous, Iceland, and Audouin's Gulls.

Ponta da Cruz is the southernmost point in Madeira and is located in the west part of Funchal. It is a good spot for seawatching and from here it is possible to see Cory's, Manx and Macaronesian Shearwaters, Bulwer's Petrel and even *Pterodroma* petrels (Zino's/Fea's). This place can be easily reached from Funchal along a promenade starting at Lido Garden that ends at Ponta da Cruz after 3km.

The Funchal gardens can be a good place to look for some of the endemic species and subspecies of passerines occurring in the island. The best ones are the Jardim Botânico and the Tropical Monte Palace. Exploring these two large gardens can produce Robin, Blackbird, Blackcap, Madeira Kinglet, Chaffinch, Island Canary and Goldfinch. It is worth keeping an eye on the sky as Plain Swifts are sometimes around. Both gardens are located within walking distance of the town centre and they can also be reached using the public transport network.

b) The coast east of Funchal

East of Funchal, the small town of Machico lies just beneath the airport. The open bay there is good for terns and gulls. Because of the persistent northerly winds it can be very difficult to see *Pterodroma* petrels from here but with a telescope turned on Bugio (in the Desertas), the extraordinary gad-fly action of birds can occasionally be seen, normally at very long range. Here also Spanish Sparrows begin to appear and Island Canaries can often be seen. The island subspecies of Buzzard and Kestrel can be found. The mouth of the small Machico river is worth looking as it can attract small parties of waders, some gulls and terns. It is worth walking upwards along the stream. The local race of Grey Wagtail and the introduced Common Waxbill are regular and rarities like Semipalmated Plover and Killdeer have already turned up there.

Past Machico and close to the village of Caniçal lies Ponta de São Lourenço, a dry and treeless headland which forms the eastern end of the island. Berthelot's Pipits, Rock Sparrows and Island Canaries abound here and a myriad of grounded trans-Saharan birds can occasionally be seen. In spring and summer nights the air is filled with the sound of Cory's Shearwater. The islets at the tip of this area are a good place for breeding seabirds and the first breeding records for the island of Macaronesian Shearwater and Roseate Tern were made there although the latter was apparently an isolated case. Occasionally Monk Seals turn up along this coast. Ponta de São Lourenço can be reached from Caniçal using the R109 that ends at a car park. From there it is possible to walk eastwards in a 'desert-scape' using the marked paths.

c) The coast west of Funchal

Some 20km west of Funchal lies the tiny coastal water body called Lagoa do Lugar de Baixo. This very small wetland was created in the beginning of the 19th century after a landslide. This is the only wetland of this kind in the entire island and it attracts several species that can be found nowhere else on Madeira. Although the number of birds usually present is not very high it includes some Coots and Moorhens which breed in the area. The odd duck can be seen, as can small parties of waders like Black-winged Stilt, Oystercatcher, Dunlin, Black-tailed Godwit and others. Night-heron can sometimes also be seen and rarities have turned up there. This place can be reached from Funchal from the motorway to Ribeira Brava and following the signs to 'Lugar de Baixo' and 'Ponta do Sol'.

A bit further west, about 50km from Funchal, Ponta do Pargo is a high headland that deserves a visit, despite the fact that some ongoing touristic development projects have altered the habitat. Plain and Pallid Swifts both occur here. Besides, the entire area is good for species like Buzzard, Sparrowhawk, Spectacled Warbler and Rock Sparrow. This headland can also attract some migrants and there are records of rare species for Madeira, like Red-footed Falcon, Great Spotted Cuckoo and Red-rumped Swallow. Ponta do Pargo can be reached from Funchal via the motorway to Ribeira Brava.

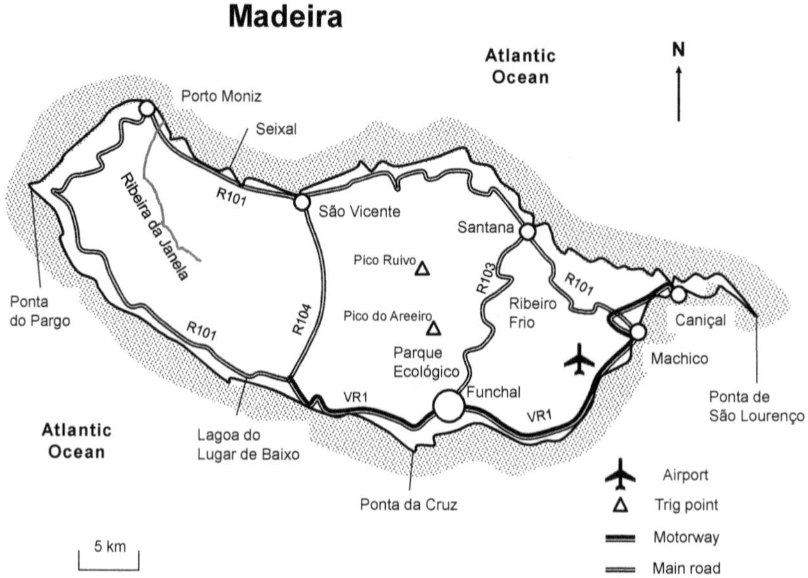

Some consider Porto Moniz on the north side of the island to be one of the best seawatching spots in the Western Palearctic. This might seem a bit exaggerated, but the truth is that watching the sea from there can be most rewarding. The number of birds passing through can be high with a considerable diversity of species. Cory's Shearwater is common and, besides, Manx, Sooty, Macaronesian and Great Shearwaters are all possible. Bulwer's Petrel is not uncommon, nor are *Pterodroma* Petrels (Zino's/Fea's). There are records of Pomarine and Great Skuas. Apart from birds, this is also a good place to look for whales and dolphins. From Funchal, the best access to the area is by the motorway to Ribeira Brava, continuing then to São Vicente and from there west to Porto Moniz. The best place to watch is close to the natural pools.

To the east of Porto Moniz and not very far from there the mouth of the small ribeira da Janela is also worth a visit. The odd duck might be around (there is one record of Green-winged Teal) and at night when the sound of the sea is not too loud it is possible to hear the calls of the Madeiran Storm Petrels that breed in the nearby rocky crevices.

d) The laurel forest and the mountains

The rugged inland of Madeira is covered with the largest stand of laurel forest in the world. Although access to some places is not easy, there are many marked paths that can be used to explore the area. Some are dangerous and not recommended to those who are afraid of heights. Parque Florestal do Ribeiro Frio some 17km from Funchal offers a good place from where to explore the laurel forest. Take the R103 north via Monte and the Poiso cross-roads at 1400m above sea level. Past Poiso, the road descends into a long deep valley to Ribeiro Frio. From the Parque Florestal it is possible to take a 30 minute walk until the viewing point of Balcões is reached. The breath-taking views over the laurisilva forest make this walk worthwhile and patience is all that is needed to find Madeira Laurel Pigeons, their tail markings unique even in the dim mist. Besides pigeons, from the viewing point it is possible to see Buzzard, Sparrowhawk and the Madeiran subspecies of Chaffinch (its calls are distinctly different to those of their continental counterparts and much of its pink coloration is suppressed). The woods here hold Madeira Kinglet.

One of the best places to see the endemic Madeira Laurel Pigeon is the Chão da Ribeira area, which can be reached from the small village named Seixal on the north coast (between Porto Moniz and São Vicente).

Parque Ecológico do Funchal (Funchal Ecological Park) is a protected area (about 1.000 ha) north of the city. Altitude ranges from 470 to 1818m. It includes some laurel forest areas at lower altitude and some patches of scrub higher up. There are several options whereby to explore it. The best one is probably the walk between Poço da Neve and Casa do Barreiro which allows the crossing of several altitude levels. Severe fires hit this part of the island in 2012 and have destroyed large portions of the laurel forest and damaged parts of the pathway. There are plans to restore it in the near future. From Funchal one must proceed to Poiso and from there proceed onwards towards Pico do Areeiro. A few kilometres after Poiso and before reaching Pico do Areeiro, there is a sign on the left to 'Poço da Neve' where the walk starts. Casa do Barreiro is reached after about 6km. Keep in mind that this walk will take 3 hours in each direction, therefore it requires careful planning and an eye for sudden weather changes.

There are other sites where one can visit the laurel forest, on the northern slopes, such as Caldeirão Verde, in Parque Florestal das Queimadas, near Santana.

The highlands of Madeira not only offer spectacular and dramatic views, they are also the only place in the world where the endangered Zino's Petrel breeds. The chances of seeing one of these protected birds are remote but night sessions to try to listen for their calls at breeding sites might be arranged. Visitors are advised to contact the Parque Natural da Madeira headquarters (see the 'Useful Contacts and Websites' section at the end of the book) or local birding tours companies in order to find out about the conditions. The walk that connects Pico do Areeiro to Pico Ruivo (both above 1800m) crosses the breeding habitat of Zino's Petrel and is quite spectacular. However it is not recommended to those who suffer from vertigo or are afraid of heights. Among the species that can be seen in this area are Sparrowhawk, Plain Swift, Berthelot's Pipit and Rock Sparrow. Pico do Areeiro can be easily reached from Poiso.

Porto Santo

Porto Santo, lying to the north east of Madeira itself is a smaller island (about 42km^2), within sight of Ponta de São Lourenço at the eastern end of Madeira and like that area, has a much drier climate. A spectacular sandy beach stretches along the south coast for about 10km. The vegetation is relatively sparse and concentrated into a few places outside Vila Baleira, the largest town. The island has a maximum height of 283m at Pico de Ana Ferreira.

Because of its drier climate, the birds that are typical of the Madeira laurel forest are absent and the bird life in Porto Santo is rather similar to that of the eastern, drier part of Madeira. Common and widespread landbirds in Porto Santo include Plain Swift, Berthelot's Pipit, Spanish and Rock Sparrows and Island Canary. Other breeding species are Hoopoe and Spectacled Warbler. The surrounding islets are home to breeding seabirds, such as Cory's and Macaronesian Shearwaters, Bulwer's Petrel and Madeiran Storm Petrel. Outside the breeding season it is possible to see many species of migrant birds, both from the Palearctic and from the Nearctic.

There are two ways of getting to Porto Santo: by airplane or by boat. In the first case, there are daily flights from Funchal which are operated by SATA; in the second case, it is possible to use the daily ferry (the 'Lobo Marinho', costing about €56 return, as of 2013, about €10 less in low season) which crosses from Funchal to Porto Santo in one hour (for timetables please refer in both cases to the 'Useful Contacts and Websites' section at the end of the book). Once on Porto Santo, it is possible to hire a car in order to travel around the island, although it is advisable to book it well in advance.

Travelling by boat from Madeira can be a good opportunity to seawatch and most of the 'special' seabirds of this region can be seen including the *Pterodroma* petrels (Zino's / Fea's). Once on Porto Santo, assuming that one has arrived by boat, the best strategy is to walk from the town westwards to Porto da Calheta, the westernmost point on the island (a walk of about 6km). The 500m wide strait called Boqueirão de Baixo, between Porto Calheta and the Ilhéu de Baixo, is worth watching at dawn as Bulwer's Petrels and Macaronesian Shearwaters leave to feed offshore (remember that in winter Bulwer's Petrel may be absent and *Pterodroma* petrels difficult to find).

There are three small water bodies on Porto Santo, at the foot of Pico de Ana Ferreira, at Lapeiras and the appropriately named Tanque, quite close to Vila Baleira itself. Nearctic waders have been seen here. In the eastern part of the island, there is a golf course which also attracts waterbirds, although permission is necessary to walk around. Altogether Porto Santo has all the makings of an excellent bird observatory, especially during migration times.

Desertas

Literally meaning 'Desert', the Desertas form a group of three islands, lying about 25 nautical miles (about 46km) south-east of Madeira.

They are all barren and uninhabited. Deserta Grande, the largest of them, lies in the middle and rises to over 400m above sea level. To the south of it lies Bugio and to the north the smaller, flat Ilhéu Chão (meaning 'ground islet'),

This small archipelago is important for its seabird populations. Breeding species include Cory's and Macaronesian Shearwaters, Bulwer's Petrel, Madeiran Storm Petrel and the rare Fea's Petrel (some authorities consider the form, which breeds on Bugio only, as a proper species endemic to these islands – the Desertas Petrel *Pterodroma*

Pterodroma

deserta). Other breeding species include Yellow-legged Gull (*atlantis* race) and Common Tern.

The diversity of landbirds is small, due to the lack of vegetation. Resident species here include Buzzard, Kestrel, Barn Owl, Plain Swift, Berthelot's Pipit and Island Canary.

Besides birds, Desertas are also home to the threatened Monk Seal, of which there are still a few individuals around, although very difficult to see.

There are no regular passenger ferries to the Desertas, so the only way to get there is on a private boat. A few companies organize regular birdwatching visits. Because this is a nature reserve, there are restrictions on access and most of these islands cannot be visited. However the boat-trip is worth taking because of the seawatching opportunities (many of the above named seabird species can be seen).

Selvagens

The Selvagens ('savage') archipelago lies some 163 nautical miles (300km) south of Madeira off the North African coast and only 82 nautical miles (150km) from the Canary Islands. It is composed of three larger islands and a few small islets. The main island is Selvagem Grande. The other two are Selvagem Pequena and the Ilhéu de Fora. The total land surface is small, not exceeding 265 ha. Its geographical features and location favour it with an altogether different ambience, almost like a wholly different set of islands attached to the Madeira archipelago. The Selvagens have a rich history involving the exploitation of Cory's Shearwater chicks for food and fat and the conservation work done by Paul Alexander Zino and others who pioneered the creation of a protected area in 1971.

The Selvagens are dry and treeless. The Selvagem Grande has an extensive plateau 100m above sea level and its highest point is Pico da Atalaia (163m). The coast is rugged and it has several small caves and bays. Selvagem Pequena is lower and the entire island is barely above sea level (the highest point is just 49m high). It has a long sandy beach, where White-faced Petrels breed. The vegetation of the Selvagens is composed mainly of scrubby xerophytic plants and is of enormous scientific interest. The flora includes at least 105 species of which 11 are endemic. There is no permanent human settlement except for rangers from Parque Natural da Madeira that stay there year round, on shift duty.

This small archipelago is a very important place for breeding seabirds. There is a large Cory's Shearwater colony. Bulwer's Petrel, Madeiran Storm Petrel and Macaronesian Shearwater also breed. There are even records of Swinhoe's Petrel and it is not impossible that this species may breed occasionally. The Selvagens are also one of the very few Western Palearctic breeding sites for White-faced Petrel. Sooty Terns can regularly be seen and the odd pair has bred in some years. In the surrounding waters there are more or less regular sights of Red-billed Tropicbirds and Cape Verde Shearwater has already been seen. Berthelot's Pipit is the only species of landbird breeding and is represented by the nominal race (different from the one occurring at Madeira and the same as the one in Canary Islands).

Given its geographical location close to the African coast, the Selvagens are visited by a vast array of Western Palearctic migratory birds especially during autumn migration. Apart from the more regular species at that time of the year like swallows or certain warblers, many rarities have been reported. Among them are for instance, Eleonora's Falcon and Red-throated Pipit. Even American vagrants like Yellow Warbler have turned up there. It is worth noting that this place is underwatched and that the potential for vagrancy is enormous.

Reaching the Selvagens is not easy and until a few years ago it would have been almost impossible for those not engaged in some sort of investigation project. Even so, it would have been necessary to go with the rangers on shift and return with them after a three-week stay. Nowadays it is possible to arrange a boat trip from Madeira. As often happens with this kind of trips, weather conditions, and especially wind strength, can be an important limiting factor concerning departures times so care is needed to schedule flights to Madeira if planning an onward journey to Selvagem (allow one or two days before and after the boat trip). The boat trip can take 12 hours or more one way and the sea can be rough (if you suffer from sea sickness, don't forget pills). For those interested in seabirds, the best time to go is probably between May and August. Before booking, make sure that the trip comprises a stay of at least two days. Once there, it is possible to visit Selvagem Grande plateau, escorted by one of the rangers.

GLOSSARY OF LANDSCAPE TERMS

Barragem. A dam, usually made of concrete.
Cabo. Headland or cape (mainland only).
Caldeira. Remains of old volcanic activity (Azores only).
Fajã. A flat area along the sea line, usually formed by lava or boulders; most fajãs lie at the base of steep cliffs (Azores and Madeira only).
Ilhéu. A small islet, usually uninhabited.
Lagoa. A coastal lagoon on sandy terrain which may or may not be connected to the sea; on the Azores and serra da Estrela: a natural lake.
Paul. A freshwater marsh, usually with reedbeds or other type of aquatic vegetation.
Pico. Highest point in a mountain (peak).
Ponta. Headland or cape.
Ria. A coastal wetland which may be separated from the sea by a chain of barrier islands; can also mean an estuary which is separated from the sea by two sandbars.
Ribeira. A small river, a stream.
Serra. A chain of hills which usually extends along 10 to 50km and which can reach a height of 500m to 2000m.

SELECTIVE BIRD LIST

Mainland Portugal

Red-crested Pochard. Breeds regularly in small numbers at a few places in southern Portugal. In winter large flocks are sometimes seen at Lagoa de Santo André. Ringing returns suggest that most of these are of Spanish origin.

Cory's Shearwater. It breeds only in the Berlengas archipelago, where it is common, mainly from May to August. It can be found however from February onwards along the entire coast, best seen from prominent headlands, like Cabos Raso and São Vicente or during pelagic seawatching trips.

Madeiran Storm Petrel. A very scarce breeder in the Berlengas archipelago, where it was first found in 1980 in one of the outer islets of the archipelago. This bird of apparently oligotrophic waters has also been heard singing over Berlenga Grande itself. This nucleus forms the only breeding population of the species in continental Europe, but, like other petrels, can be very difficult to observe due to its nocturnal habits.

Little Bittern. A scarce summer visitor to some reedbeds and to dense riparian cover. Not always easy to observe. Among the places to look for it are at the Lower Mondego marshes, Paul de Tornada, Lagoa de Santo André and Quinta do Lago, although it is known from a variety of other sites in central and southern Portugal. The 'song' is distinctive and dawn and dusk are the optimum times to search for it.

Night-heron. This migrant heron has a very localised distribution in Portugal. There is a large colony at Paul do Boquilobo and a smaller one in Escaroupim, near Salvaterra de Magos. Odd individuals turn up at other sites.

Squacco Heron. A heron of uncertain status, which has bred irregularly at Paul do Boquilobo and Escaroupim in very small numbers. The most reliable place to look for it is however the Paul da Barroca where some birds have been seen regularly in the last few years. Sightings are occasionally made at other sites throughout the year – a few overwinter.

Great White Egret. Formerly a vagrant, but now a regular non-breeding visitor in small numbers to most large wetlands. Can be regularly seen at the Lower Mondego marshes, Paul da Barroca, Tagus and Sado estuaries and Lagoa de Santo André, as well as in some reservoirs in the eastern Alentejo.

Purple Heron. A summer visitor to many wetlands in the country, where it is locally common and may form small colonies in large reedbeds. More numerous in coastal areas, such as Ria de Aveiro, the Lower Mondego marshes and the Tagus and Sado estuaries, it breeds also in certain inland sites, like Paul do Boquilobo.

Black Stork. A rare breeding species, restricted to relatively undisturbed, mostly uninhabited inland regions. The important sites are at International Douro and International Tagus. Smaller numbers breed in the eastern Alentejo, namely along the Guadiana valley and around Barrancos. During migration it is regularly seen at Cabo de São Vicente in autumn and it can be seen occasionally in other coastal areas. A small number of birds may overwinter in the south.

Glossy Ibis. Twenty years ago it was a rare visitor to Portugal, but in recent years it has been recorded in good numbers in coastal wetlands, such as the Tagus and the Sado estuaries and the Paul da Barroca, mostly outside the breeding season. It has bred at Escaroupim and at Paul do Boquilobo.

Spoonbill. Mainly a passage migrant and wintering species, which can occur in any wetland, although it is usually most numerous at Castro Marim, Tagus and Sado estuaries where flocks of over 100 birds are regularly seen. In 1988, this bird was found breeding in Portugal for the first time in the 20th century, at Paul do Boquilobo, and since then its breeding has been recorded at a few other sites.

Greater Flamingo. The species does not breed regularly, but is usually present throughout the year in the Ria de Aveiro, Tagus and Sado estuaries, Ria Formosa and Castro Marim areas, with several thousand birds at certain times of the year, especially in autumn. Can turn up elsewhere in smaller numbers including small to medium size reservoirs inland.

Black-winged Kite. This resident raptor has increased in recent years and is widely distributed over most of southern and eastern Portugal. It is easy to observe in the Tagus estuary and is fairly common in the province of Alentejo, where it can be seen at most sites mentioned in this book.

Red Kite. A decreasing breeding species, nowadays very scarce and restricted to the more remote inland regions close to the Spanish border north of the Tagus, mainly in the north-east, such as International Douro. More numerous and widely distributed in winter, namely in the Alentejo plains.

Egyptian Vulture. As a summer visitor it breeds in the eastern part of the country, but is usually scarce, except at International Douro, where it is still reasonably common. It can also be found at the International Tagus, but it no longer breeds over most of the Alentejo. Occurs regularly on autumn migration at south-western watch points like Cabo de São Vicente.

Griffon Vulture. Locally common in certain inland regions, where it can be seen in small flocks. The largest numbers can be seen in the north-east at International Douro and also at International Tagus and Portas de Ródão. Further south it is also seen regularly at the serra de São Mamede, Barrancos and Mértola areas. Rather scarce elsewhere, although it usually occurs yearly in good numbers on autumn migration at Cabo de São Vicente.

Rueppell's Vulture. Although still considered a rarity, this species is recorded almost annually at Cabo de São Vicente where isolated birds turn up among the big vulture flocks in autumn. Furthermore, one or two birds have been recorded at Portas de Ródão. There was even a breeding attempt there between a mixed pair of vultures: a Rueppell's and a Griffon.

Cinereous Vulture. In the last years there have been some isolated breeding attempts near the Spanish border. It is usually seen in small numbers, most often singly among flocks of Griffon Vultures. Occurs mainly in the International Tagus, Barrancos and Mértola. Each autumn a few isolated birds are reported at Cabo de São Vicente. Records at other sites are very rare.

Short-toed Eagle. A widespread summer visitor (as Booted) arriving in March, found over most of the country but most abundantly in the south. Serra do Gerês, Mora and Montargil and the Tagus estuary are good sites to find it. Birds on passage occur at Cabo de São Vicente in autumn.

Spanish Imperial Eagle. After many years of absence, this large eagle has recently started to breed once again: a few pairs have been found nesting in the Alentejo and Beira Baixa provinces. Young birds occasionally turn up elsewhere.

Booted Eagle. A widespread summer visitor, it occurs mainly in the southern half of the country, where it can be locally common. Possibly the best region to locate it is the Alentejo, especially in open Cork Oak woods. Numerous in autumn near the Cabo de São Vicente. Booted Eagles overwinter in very small numbers in the southern half of the country.

Bonelli's Eagle. An uncommon resident species with a very localised distribution. Occurs mostly in remote areas and in very small numbers. Regularly sighted at International Douro, International Tagus and serra da Arrábida, among other places. Many pairs breed in cliffs or rocky outcrops but in the south-west there is a tree-breeding population.

Lesser Kestrel. As is the case in many European countries, this small falcon has declined sharply in the late 20th century, but in the last few years its population has recovered to some extent, with the largest numbers now found around Castro Verde and Mértola. It is also recorded occasionally as a passage migrant at various places, mainly coastal sites, usually in autumn. Kestrel, which often breeds colonially, can cause identification problems.

Eleonora's Falcon. A very scarce passage migrant, mainly in late summer and autumn. Most sightings are from the Cabo de São Vicente area. Breeding has been suspected at the rocky coast north of Lisbon but never proved.

Baillon's Crake. Information about this species is very scarce and its status is uncertain. It is assumed to be a very rare summer visitor or possibly a passage migrant.

Purple Swamphen. Formerly an extremely rare breeder, it is nowadays to be found in most wetlands in southern and central Portugal. It is a common bird in central Algarve and can be easily seen at Quinta do Lago. Lagoa de Albufeira, not far from Lisbon, is also a good place to observe it.

Red-knobbed Coot. A few records from the Algarve (Ria Formosa and Quinta do Lago), western Alentejo (Lagoa de Santo André) and Tagus estuary suggest that this species occurs more regularly than previously thought; a recent reintroduction program in Spain may have helped increase the number of sightings.

Common Crane. A winter visitor with a localised distribution in the Alentejo. Occurs regularly from November to February around Elvas, Castro Verde, Mourão and Barrancos, where it feeds both in open Holm Oak groves and cereal fields. It can be difficult to locate and should definitely not be disturbed.

Little Bustard. A fairly common resident breeder in southern Portugal, which also occurs further to the north, inhabiting mainly large open fields. Large flocks can be found in winter. May be decreasing, though it can still be found relatively easily around Castro Verde, Mourão and Elvas. Close to Lisbon it occurs in small numbers in the Tagus estuary during the winter.

Great Bustard. Has decreased in recent years, due to habitat changes, but still occurs locally in small numbers throughout the Alentejo. Most regularly found on the Castro Verde plains, after harvesting and before spring growth of cereals conceals them again.

Stone-curlew. Dusk singing during the breeding season is often the first sign of its presence. At that time of year this resident species is widespread and occurs in both steppe areas and open oak woods. In winter, it tends to flock and is always difficult to detect and ascertain numbers. It seems to be more common in the province of Alentejo.

Collared Pratincole. A local summer visitor to southern areas, breeding in open fields and fallow land, often close to water bodies. It occurs regularly at the Tagus estuary, Elvas, Castro Marim and Évora; scarce on passage.

Audouin's Gull. Following a population increase in Spain, this gull is now recorded annually, mainly in the Algarve and it started breeding there from 2000 onwards. Most records outside the known breeding grounds are from late summer, although a few birds can also be seen in winter.

Caspian Tern. A scarce but regular winter visitor in small numbers to wetlands on the south coast (Rias de Alvor and Formosa and also at Castro Marim); up to twenty have been seen together at coastal roosts. Rather scarce elsewhere, mainly as a passage migrant and, occasionally, in mid-summer, but a few winter regularly in Sado estuary. Ria Formosa may be one of the best places to find this species in Europe during the winter.

Whiskered Tern. An irregular breeder, its numbers and distribution are much dependent on winter rainfall and the resultant fluctuations in the water table in certain areas. Has bred at Paul do Boquilobo and at a few places in the Alentejo. A few non breeders often turn up in early summer in the rice fields near the Tagus estuary. This tern is also a scarce passage migrant and odd individuals have been seen in winter.

Black-bellied Sandgrouse. Nowadays a very rare resident, probably due to changes in agricultural practices. Used to be seen regularly between Castro Verde and Mértola, but sightings have decreased in recent years. Still occurs regularly at Mértola, Mourão and Évora, mainly in steppe-like habitat.

Pin-tailed Sandgrouse. Apparently once common, it is now an extremely rare species in Portugal. A small resident population is known from the International Tagus region.

Great Spotted Cuckoo. A scarce summer visitor, which breeds locally. It is nowhere common, but seems to be slightly more frequent in the eastern part of the country. The best times to watch it (and hear it) are late February and March (when adult birds are displaying) and June (when juveniles start to fly). Otherwise it is a shy bird which can be extremely difficult to locate; scarce on passage.

Scops Owl. A summer visitor, mainly to the northernmost part of the country (serras do Gerês and Montesinho, International Douro and Ribacoa), where it is common, sometimes singing in the middle of town parks. Rather scarce in southern Portugal as a breeding bird, although it occurs regularly in the Algarve on autumn migration. Like most owls, more often heard than seen.

Eagle Owl. A rare resident, which occurs extremely locally throughout the country, but mainly in remote inland areas, like the International Douro, International Tagus and Barrancos areas. Usually it is difficult to locate, due to its nocturnal habits, though the distinctive song may be the first and only sign of its presence.

Red-necked Nightjar. A fairly common summer visitor, except to the north-western part of the country. Occurs in a variety of habitats, but most often in a mixed landscape of woods and dry open fields. Best located from late April to early July, when birds sing at dusk.

Pallid Swift. A common summer visitor mainly to the southern half of the country, although it occurs up to Ria de Aveiro along the west coast and as far north as Bragança in the north-east. It is much overlooked because of confusion with Common Swift. There are sizeable colonies in the cities of Aveiro, Lisbon, Setúbal, Elvas and Faro, and it occurs regularly at Cabo de São Vicente, at serra da Arrábida and at the Tagus and Sado estuaries.

Alpine Swift. A common summer visitor to the International Douro and south-west coast areas, which also breeds locally in other places, such as the western Algarve. Occurs regularly as a passage migrant over most of the country.

White-rumped Swift. As in Spain, there has been a recent increase in reports of this species, mainly from the southern half of the country though it is still very rare. Breeding has been proved near Mértola and at a few other places, mostly in the south-east.

Roller. An uncommon summer visitor to open areas in the eastern part of the country. It occurs at Castro Verde, Mourão and a few other places in the Alentejo, usually in low densities. On autumn migration, it is regularly seen at Cabo de São Vicente and sometimes in other places like the Ponta da Erva plains in Tagus estuary or Cabo Espichel.

Wryneck. An uncommon summer visitor and breeding species, which is most easily detected in April and May by its song. Best looked for in the north-eastern part of Portugal, like serra de Montesinho and International Douro, where it occurs in open habitats with some trees. It is regularly observed as a passage migrant in autumn and individuals are occasionally recorded in winter in the southern half of the country.

Calandra Lark. A resident bird, which breeds in open, steppe-like areas. Occurs mainly at International Tagus, Évora, Mourão and Castro Verde, although it can be seen regularly in other places with suitable habitat.

Short-toed Lark. A summer visitor which can be locally common, breeding mainly in dry open habitats, most numerous in the southern half of the country.

Lesser Short-toed Lark. A small population occurs in the Castro Marim area, where it breeds in dry saltmarsh and is present throughout the year.

Thekla Lark. A common resident, more widespread in the eastern half of the country. It seems to prefer rocky, uncultivated terrain, sometimes with few trees. Best looked for in the Ribacoa, International Tagus, serra de São Mamede, Castro Verde and Mértola areas, but it is quite common from Cabo de São Vicente north to Cabo Sardão as well. In the Elvas area, both 'crested' larks are common, allowing for comparison.

Crag Martin. A resident breeder in the eastern half of the country, which occurs mainly in river valleys. Look for them near large dams, rocky outcrops, stone bridges and old castles. In winter, it has a more widespread distribution and occurs in wetlands like Lagoa de Albufeira and coastal cliffs, where it may form large roosts.

Red-rumped Swallow. A summer visitor breeding throughout the country. It is most numerous in the south. It nests under bridges, in old castles and ruins.

Tawny Pipit. A summer visitor with a localised distribution. It occurs mainly in the northern hills (e.g. serras do Gerês, Montesinho and Estrela), where it is fairly common. In the Alentejo plains it is generally a scarce bird, although it can be locally common, such as in the Évora region or along the coastline. Occurs as a passage migrant in coastal areas, especially in autumn.

Alpine Accentor. A very scarce winter visitor with an extremely localised distribution; regular sightings however occur at Cabo de São Vicente lighthouse, serras da Estrela and Sintra, at the rocky bluffs on the upper Tagus (notably at Portas de Ródão) and some of the castles of eastern Portugal (e.g. Marvão, near serra de São Mamede).

Rufous Bush Chat. A scarce summer visitor, which prefers riversides with Sun Rose and Oleander. It is restricted to the south-eastern part of the country. A late breeder, arriving in May, it should be looked for around Castro Verde, Mértola and Castro Marim. Can also be found at Reguengos de Monsaraz, near Mourão; rather rare on passage.

Bluethroat. Common but localised passage migrant and winter visitor to coastal wetlands, such as Ria de Aveiro, Cávado, Tagus and Sado estuaries and the Ria Formosa area.

Black-eared Wheatear. A summer visitor, which is most common in the Alentejo (Évora, Mourão, Castro Verde, Mértola, and Barrancos) and the International Tagus but is rather scarce elsewhere, although it can be found as far north as serra de Montesinho. It occurs in a variety of habitats, mainly in open country.

Black Wheatear. A resident species, largely restricted to the international Douro and the International Tagus.

Rock Thrush. Breeds in the high mountains of the northern half of the country, but the breeding density is very low indeed. The best places to look for it are serras do Gerês, Montesinho and Estrela. Occasionally recorded on passage at coastal sites.

Blue Rock Thrush. A resident species, most commonly found in deep rocky river valleys, and along rocky coasts, but also to be found locally on inland castles. Can be most easily located on the steep slopes of places like serra da Arrábida and Cabo de São Vicente, where it is seen at sea level, and in remote inland areas, like International Douro, Ribacoa, International Tagus, serra de São Mamede and Barrancos.

Savi's Warbler. A rather localised summer visitor to some reedbeds. It is common at Ria de Aveiro, Lower Mondego marshes and Lagoa de Santo André and is known to occur at Minho estuary and Paul de Tornada, but is rather scarce elsewhere. Rarely seen on migration.

Western Olivaceous Warbler. A rare summer visitor, largely confined to the south-easternmost regions. Due to its rarity and to the lack of information available, it is not possible to indicate specific sites. However, there have been some records from the area along the river Guadiana south of Mértola.

Spectacled Warbler. An uncommon breeding species with a localised distribution. Although it is a summer migrant, it arrives very early, usually in late February or early March. Best looked for east of Mértola (where it occurs in low scrub) or in Castro Marim (where it breeds in saltmarshes). It is also known from Freixo de Espada à Cinta, in the International Douro, the Tagus estuary and Cabo de São Vicente.

Subalpine Warbler. A common summer visitor to the north-eastern part of the country, where, remarkably, it may be the most numerous warbler in bushy areas (e.g. serra de Montesinho, International Douro and Ribacoa). It can also be found further south, for example at serra da Gardunha. Much less common in the eastern Alentejo, where it can be found, mainly in Sun Rose scrub, near Mourão or Mértola. Not uncommon on passage, often seen as early as mid-March in breeding grounds.

Western Orphean Warbler. Uncommon as a breeding species over most of the country. Its distribution is patchy and it is unlikely to be seen in most places. It can, however, be found regularly around International Douro, Ribacoa and International Tagus, mainly in open woods, and in valleys north of Tavira and Castro Marim, as far north as Mértola.

Western Bonelli's Warbler. A summer visitor which occurs mainly in the northern part of the country (serras do Gerês and Montesinho, International Douro, Ribacoa and serra da Estrela), where it is rather numerous in Pyrenean Oak and Maritime/Scots Pine woods although it can occur in other habitats as well. In the south it is most often seen in Cork Oak or Maritime Pine woods, with understorey, especially in the Ribatejo and in the western Alentejo (namely around Mora and Montargil); occurs regularly on autumn migration.

Iberian Chiffchaff. Formerly considered a subspecies of Chiffchaff, but now widely recognised as a full species. This is a widespread summer visitor with a patchy distribution over the country. Tends to favour places with some humidity, such as riparian forests in lowlands but can also occur in dense oak stands. Good places to look for it are serra do Gerês, Mora and Montargil, and serra da Arrábida. Its song can be heard from late February onwards.

Penduline Tit. A winter visitor to reedbeds, occurring locally over most of southern Portugal. If looked for in suitable habitat, can be found in Tagus and Sado estuaries, at Lagoas de Santo André and Albufeira and also at several wetlands in the Algarve. Has bred near Elvas.

Red-backed Shrike. A late summer visitor (arriving mid-May) to the northernmost part of the country. In serra do Gerês it is common and occurs in open plateaus throughout the park, whereas in serra de Montesinho it is rather uncommon. Not usually recorded outside its breeding grounds.

Azure-winged Magpie. A locally common resident, which in Europe is restricted to central and southern Iberia. Can be found in most places south of the river Tagus and also at a few places further north, mostly along the Spanish border. It is particularly numerous in the Ribacoa, International Tagus and Mértola areas, as well as in the central part of the Algarve.

Chough. A resident species which in Portugal is restricted to four sites: Cabo de São Vicente, where it breeds in coastal rocks; serras de Aire e Candeeiros (near Leiria, in the central part of the country), where it breeds in vertical caves; serra do Gerês, where it occurs in rocky terrain; and in the International Douro, where it occurs in the rocky valley of the river and those of its tributaries; nowhere common.

Spanish Sparrow. A common summer visitor to the International Tagus region and over most of the Alentejo province. Nests can be found in road side trees, on telegraph poles and also in White Stork's nests. In winter can be seen near Ponta da Erva in Tagus estuary and at a few other coastal wetlands.

Rock Sparrow. A widespread resident with a localised distribution. Occurs in a variety of habitats, from rocky valleys to Cork Oak woods. May be seen at times in Sand Martin colonies. Rather frequent in the woods around the Tagus and Sado estuaries, but present in most of the sites described in this book.

Rock Bunting. A resident species which is very common in the northern provinces, especially in wild hilly country, and relatively uncommon and localised in the southern half of Portugal, where it is restricted to some small rocky valleys. The best sites to look for it are undoubtedly serras do Gerês, Montesinho, Estrela and Gardunha and Portas de Ródão.

Ortolan Bunting. A summer visitor to the higher mountains in the northern part of the country. It breeds in serras do Gerês, Montesinho,

Gardunha and Estrela. In the southern part of Portugal, it occurs mainly on migration, regularly in autumn and, less frequently, in spring.

Azores

Black Duck. This Nearctic duck has been reported regularly over the last few years and it was found breeding at Flores in 2008 and 2011. So far, these are the only confirmed records. Besides these occurrences, there are reports of some other breeding attempts and of some cases of hybridization with Mallards at Flores and Corvo. In autumn and winter, vagrants are sometimes reported.

Fea's Petrel. There is a record at Ilhéu da Praia close to Graciosa of a bird with a brood patch in 1999. In the last years there have been several records that suggest the existence of a small breeding population in the Azores, but the information about this species is scarce.

Bulwer's Petrel. This is a rare spring and summer visitor which has a very small breeding population at some islets off Graciosa and Santa Maria. The best chances of finding it are during pelagic trips.

Cory's Shearwater. It is the most common seabird throughout the archipelago. In fact, Azores holds probably the largest breeding population in the world. It can be seen not only during land-based seawatches but also in pelagic and inter-islands crossings. At night their strange calls fill the air close to the breeding colonies that are scattered all over the rocky shores and islets of all nine islands.

Macaronesian Shearwater. Breeds in all the islands of the archipelago with the possible exception of Terceira. It is a winter breeder but most of them remain in Azorean waters all year round. It is usually very difficult to see, as it spends the day at sea, only returning to the breeding grounds at night. The best chance to observe it is during pelagic trips or while travelling by boat between islands.

Madeiran Storm Petrel. It breeds from October to December in small islets off Graciosa and Santa Maria. It is rare and difficult to see. Like the previous species, it spends the day at sea only returning to breeding grounds at night. The best chances to get it are during pelagic trips or during inter-island crossings. Note however that identification at sea can be very difficult, as it is almost impossible to separate it visually from the Monteiro's Storm Petrel.

Monteiro's Storm Petrel. The acceptance of this taxon as a full species is not consensual yet and some authorities consider it as a race of the Madeiran Storm Petrel. It breeds from May to September in small islets off Graciosa. There are also acoustic records at Flores and Corvo but so far breeding has not been confirmed on these islands. It is rare and hard to see and very difficult to separate from the similar-looking Madeiran Storm Petrel.

Moorhen. Resident but not common. It has a very localised distribution. Breeds from March to August in lakes and marshes at São Miguel, Santa Maria, Terceira and Flores. Local race *correiana* is not recognised by most taxonomic authorities. Records in autumn and winter outside the known breeding range may refer to vagrants from continental Europe.

Killdeer. In 2010 and 2011 there were reports of breeding attempts in Santa Maria, but the information provided is scarce and the exact status of the species in the region cannot be ascertained.

Kentish Plover. This is a rare and localised resident. It only breeds at Santa Maria and Terceira (has bred at Graciosa). Outside the breeding season there are records at the central group of islands which may refer to vagrants.

Woodcock. This elusive wader is a scarce resident. It breeds on all the islands with the exceptions of Santa Maria and Graciosa. It is more numerous in the larger islands of the central group and also in Flores.

Roseate Tern. This spring and summer visitor breeds on all the Azorean islands, sometimes forming mix colonies with Common Terns. Among the best places to look for it are Ponta Delgada harbour and Caloura, at São Miguel; Praia da Vitória fishing harbour and Ilhéu das Contendas, at Terceira; Santo António and Ponta dos Ouriços, at Pico; Ilhéu do Topo, at São Jorge; Capelinhos, at Faial; the islets off Graciosa and Santa Maria. There is also a large colony at Flores, not far from the airport.

Sooty Tern. Very rare spring and summer visitor. Breeding has been confirmed at the Ilhéu da Vila off Santa Maria and occasionally at Graciosa. However there are also isolated records at São Jorge, Faial and Pico.

Long-eared Owl. The only Azorean nocturnal raptor. It is a resident species and as elsewhere it is very hard to see, although it might be more common than the existing records suggest. It does not breed in the islands of the western group although there are odd records there but it can be found on all other islands.

Goldcrest. This small bird occurs on all the Azorean islands except Corvo. It is more or less common throughout its range, the only exception being on Graciosa where its status is uncertain. The species is represented in the Azores by three endemic subspecies: the race *azoricus* in São Miguel; the race *sanctae-mariae* in Santa Maria; and *inermis* in the central and western groups.

Island Canary. Resident and common, occurring on all islands. It tends to prefer more open country landscapes with pastures, agricultural fields and coastal scrubs.

Azores Bullfinch. It is the only passerine endemic to the archipelago (a few taxonomic authorities still consider it a subspecies of Bullfinch). It can only be found in high altitude laurel forest patches in eastern São Miguel, namely at serra da Tronqueira and Pico da Vara.

Madeira

Zino's Petrel. A rare and endangered endemic breeding bird, which nests in the high mountains of central Madeira. Its population has been recovering slowly but remains at a critical level (a few dozen pairs). It only comes to land at night but can sometimes be seen at sea, although identification can be problematic, due to confusion with the following species.

Fea's Petrel. This is a Macaronesian endemic, which as a breeding species is restricted to Madeira, Cape Verde and, possibly, the Azores.

The Madeiran population breeds in Desertas, in an access-restricted area, so the best chances to get in visual contact with it is at sea or, with luck, during seawatching from land. Some authors consider the form breeding at Desertas as a full species, the Desertas Petrel *Pterodroma deserta*.

Bulwer's Petrel. Breeds in moderate numbers on islets off eastern Madeira and in much larger numbers on the Desertas, Porto Santo and especially Selvagem Grande. Summer and autumn observations are regular off Funchal and Ponta da Cruz, just west of Funchal. This species is not usually difficult to watch at sea during pelagic trips, albeit in small numbers.

Cory's Shearwater. Breeds on Madeira as well as on the Desertas and on Porto Santo. The largest population in the archipelago is on Selvagem Grande where about 15,000 pairs breed annually. The species can be seen all year round from many of the mainland's headlands and is often heard calling at night. It is also relatively easy to find at sea, both during pelagic trips or when seawatching from land.

Macaronesian Shearwater. Breeds in small numbers on islets off eastern Madeira, the Desertas and Porto Santo, although accurate counts are not available. On Selvagem Grande about 4,000 pairs breed annually. Like other shearwaters, it only comes to land at night, so the best chances to watch it are during seawatching sessions or on pelagic trips. Small numbers are seen from Ponta da Cruz, just west of Funchal and a few are often found when sailing to Porto Santo and back, especially towards dusk. The best period seems to be June to September, when young leave the nest for nearby waters.

White-faced Petrel. The Selvagens hold the only breeding birds in the Madeira archipelago with about 60,000 pairs. Present from January onwards, and probably dispersing in late summer; the extent of its movements is not well known, as it is not often seen away from these islands, but has been reported occasionally in the vicinity of Madeira.

Madeiran Storm Petrel. This small seabird has important colonies at Desertas and Selvagens, whereas in Madeira and Porto Santo it is much less common. It only comes to land at night and is difficult to spot at sea due to its small size so patience is required to find this species. Sometimes it is seen from boats plying between Funchal and Porto Santo, especially between June and September.

Kentish Plover. Uncommon and local resident which breeds only in Porto Santo.

Roseate Tern. An irregular breeding species which has nested in certain years in the Funchal area and also near São Vicente, in the northern part of the island.

Sooty Tern. A breeding visitor to the Selvagens, where one pair has been found breeding in successive years. Very rarely seen elsewhere.

Madeira Laurel Pigeon. About 7,000 of these unique birds exist on Madeira island, their main habitat being the extensive laurel forest. The best places to observe the species are steep-sided forest-clad valleys. Typical sites include Fajã da Nogueira and Balcões/Ribeiro Frio on the ER103. It also occurs in north-western Madeira around Vale da Ribeira da Janela, especially at Rabaçal on the ER204.

Plain Swift. A Macaronesian species, this small swift is a common breeder which can be easily seen throughout Madeira and Porto Santo. It breeds in rocky clefts of the many 'ribeiras', along the road-cuttings east of Funchal and at least one sea cliff site near Câmara de Lobos, west of Funchal. Little is known about its migratory habits, but the species is largely absent during winter months and it is believed that most birds move away towards north-west Africa. It appears not to occur on the Desertas or Selvagens islands.

Pallid Swift. A scarce summer visitor which can be seen in small numbers along the south coast of Madeira. The current status of this species remains partly unclear, due to problems in its identification.

Berthelot's Pipit. A common resident which can be found in open and dry areas throughout the archipelago. The easiest places to find it are Ponta de São Lourenço in Madeira or anywhere on Porto Santo. It also occurs in the Selvagens (where it is the only breeding land bird and is represented by a different subspecies) and in the Desertas.

Spectacled Warbler. This species (race *bella*) is resident in small numbers, it breeds in Madeira and Porto Santo. Look for it east of Caniçal or in the higher plateau called Paul da Serra.

Madeira Kinglet. Currently given full species status following a recent split from the Firecrest, this small passerine is a common resident. It is largely found in laurel forest, a habitat it shares with Madeira Laurel Pigeon and Chaffinch. Can be found over most of the island, except in the dry east. Does not occur elsewhere.

Spanish Sparrow. Reportedly arrived in Madeira in 1935 from North Africa, following strong easterlies and established itself as a breeding species. Nowadays it is a localised resident which is apparently more numerous in Porto Santo. Can also be seen in certain places in the eastern end of Madeira.

Rock Sparrow. An uncommon and local resident, which tends to favour dry and open areas. The most reliable site to locate it is at the extreme eastern tip of Madeira, just west of Ponta de São Lourenço. It also occurs on Porto Santo.

Island Canary. Limited to Macaronesian islands in its wild state, this close relative of the Serin (which does not occur on Madeira) frequents a wide variety of habitat and occurs over a wide range of altitude. Can also be found on Porto Santo and in the Desertas.

Non-native Birds

Crested Myna. An Asian species which was first seen in the wild in the late 1990s. Breeds at several locations in and around Lisbon. Good spots to look for it are Carcavelos beach, on the Costa do Estoril, and Corroios, near the Tagus estuary.

Common Waxbill. This African species was introduced in the late 1960s and is now widespread over most of the country. It is particularly common in wetland areas, including estuaries, riversides, reedbeds and lagoons with dense cover along the shores. Can also be found locally in São Miguel and Terceira (Azores) and on Madeira.

Red Avadavat. Another introduced bird (from Asia), which used to be common in the Elvas area, along riverine vegetation or in reeds. It has also been recorded in the Tagus and Sado estuaries, but its present status in these areas is unknown.

Black-headed Weaver. Breeds in reedbeds and can be locally common in places like the Paul de Tornada and Paul da Barroca.

Yellow-crowned Bishop. An African bird which is now widespread in several coastal wetlands. It is particularly numerous in the Tagus estuary, where flocks of over one hundred birds have been reported, but sightings are known from many other places in the country.

Tricoloured Munia. Has been found breeding at Paul da Barroca and also on the northern side of the Sado estuary, however there have been fewer sightings in recent years so the present situation of this Asian finch is not very clear.

FULL BIRD LIST

The list contains all known records of birds from mainland Portugal, Azores and Madeira up to December 2013. A few records are undocumented, others are from the 19th century, known only from anecdotal evidence. Between 1987 and 1994, all rarities were verified by the Iberian Rarities Committee (IRC) and since January 1995 a Portuguese Rarities Committee (PRC) has been in operation. Some records prior to 1987 which are undocumented or doubtful are currently under review. A list of species requiring verification by the PRC is available at the PRC website (please refer to the 'Useful Contacts and Websites' section at the end). The main sources of information for this list are included in the section 'Recommended Reading' at the end of the book.

The approach to taxonomy and systematics largely follows the guidelines provided by the Association of European Records and Rarities Committees (AERC). As to the English vernacular names, the list presented below follows largely the British Ornithologists' Union (BOU) recommendations. In the case of birds that are not part of the British list, the English name corresponds to that used by BirdLife International.

R/r – resident and regular breeder (common/scarce)
S/s – migrant breeder, present only in breeding period (common/scarce)
P/p – passage migrant (common/scarce)
W/w – winter visitor (common/scarce)
V – vagrant
Ext – extinct

SPECIES	Mainland Portugal	Azores	Madeira
Fulvous Whistling-duck *Dendrocygna bicolor*	-	V	-
Bewick's Swan *Cygnus columbianus*	V	-	-
Whooper Swan *Cygnus cygnus*	V	-	-
Bean Goose *Anser fabalis*	V	V	V
Pink-footed Goose *Anser brachyrhynchus*	V	V	V
White-fronted Goose *Anser albifrons*	V	V	V
Greylag Goose *Anser anser*	w	V	V
Snow Goose *Anser caerulescens*	-	V	-
Barnacle Goose *Branta leucopsis*	V	V	-
Brent Goose *Branta bernicla*	w	V	V
Ruddy Shelduck *Tadorna ferruginea*	V	V	V
Shelduck *Tadorna tadorna*	r w	V	V
Wood Duck *Aix sponsa*	-	V	V
Wigeon *Anas penelope*	W	w	V
American Wigeon *Anas americana*	V	V	V
Gadwall *Anas strepera*	r w	V	V
Teal *Anas crecca*	W	w	V
Green-winged Teal *Anas carolinensis*	V	V	V
Mallard *Anas platyrhynchos*	R W	w	P
Black Duck *Anas rubripes*	-	V	-
Pintail *Anas acuta*	w	V	V
Garganey *Anas querquedula*	s p	V	V
Blue-winged Teal *Anas discors*	V	V	V
Shoveler *Anas clypeata*	r W	V	V

SPECIES	Mainland Portugal	Azores	Madeira
Marbled Teal *Marmaronetta angustirostris*	V	-	V
Red-crested Pochard *Netta rufina*	r w	-	-
Pochard *Aythya ferina*	r w	V	V
Ring-necked Duck *Aythya collaris*	V	V	V
Ferruginous Duck *Aythya nyroca*	w	V	-
Tufted Duck *Aythya fuligula*	w	V	V
Scaup *Aythya marila*	w	V	V
Lesser Scaup *Aythya affinis*	V	V	-
Eider *Somateria mollissima*	V	V	-
King Eider *Somateria spectabilis*	-	V	-
Long-tailed Duck *Clangula hyemalis*	V	V	V
Common Scoter *Melanitta nigra*	w	V	V
Surf Scoter *Melanitta perspicillata*	V	V	V
Velvet Scoter *Melanitta fusca*	V	-	-
Bufflehead *Bucephala albeola*	V	V	-
Goldeneye *Bucephala clangula*	V	V	V
Hooded Merganser *Lophodytes cucullatus*	-	V	-
Smew *Mergellus albellus*	V	-	-
Red-breasted Merganser *Mergus serrator*	w	V	V
Goosander *Mergus merganser*	V	V	-
White-headed Duck *Oxyura leucocephala*	V	-	-
Capercaillie *Tetrao urogallus*	Ext	-	-
Red-legged Partridge *Alectoris rufa*	R	-	-
Grey Partridge *Perdix perdix*	Ext	-	-
Quail *Coturnix coturnix*	s w	r	r
Red-throated Diver *Gavia stellata*	w p	V	-
Black-throated Diver *Gavia arctica*	V	V	-
Great Northern Diver *Gavia immer*	w p	V	V
Pied-billed Grebe *Podilymbus podiceps*	-	V	-
Little Grebe *Tachybaptus ruficollis*	r	V	V
Great Crested Grebe *Podiceps cristatus*	r p w	V	-
Red-necked Grebe *Podiceps grisegena*	-	V	-
Slavonian Grebe *Podiceps auritus*	V	V	V
Black-necked Grebe *Podiceps nigricollis*	w	V	V
Black-browed Albatross *Thalassarche melanophris*	V	V	-
Wandering Albatross *Diomedea exulans*	V	-	-
Fulmar *Fulmarus glacialis*	V	V	V
Trindade Petrel *Pterodroma arminjoniana*	-	V	-
Zino's Petrel *Pterodroma madeira*	-	-	r
Fea's Petrel *Pterodroma feae*	-	s	r
Zino's/Fea's Petrel *Pterodroma feae/madeira*	V	-	-
Bermuda Petrel *Pterodroma cahow*	-	V	-
Capped Petrel *Pterodroma hasitata*	-	V	-
Bulwer's Petrel *Bulweria bulwerii*	V	s	s
Cory's Shearwater *Calonectris diomedea*	S P	S	R
Great Shearwater *Puffinus gravis*	p	p	p
Sooty Shearwater *Puffinus griseus*	p	V	p
Manx Shearwater *Puffinus puffinus*	w p	s	r p

SPECIES	Mainland Portugal	Azores	Madeira
Balearic Shearwater *Puffinus mauretanicus*	W P	V	V
Macaronesian Shearwater *Puffinus baroli*	V	R	r
Wilson's Petrel *Oceanites oceanicus*	V	V	V
White-faced Petrel *Pelagodroma marina*	V	V	R
Storm Petrel *Hydrobates pelagicus*	w p	-	V
Leach's Petrel *Oceanodroma leucorhoa*	w	V	V
Swinhoe's Petrel *Oceanodroma monorhis*	V	V	V
Madeiran Storm Petrel *Oceanodroma castro*	r	s w	r s
Red-billed Tropicbird *Phaethon aethereus*	V	V	V
White-tailed Tropicbird *Phaethon lepturus*	-	V	-
Masked Booby *Sula dactylatra*	-	V	-
Brown Booby *Sula leucogaster*	V	V	V
Gannet *Morus bassanus*	W P	p	w p
Cormorant *Phalacrocorax carbo*	W P	V	V
Double-crested Cormorant *Phalacrocorax auritus*	-	V	-
Shag *Phalacrocorax aristotelis*	r	-	V
Magnificent Frigatebird *Fregata magnificens*	-	V	-
Bittern *Botaurus stellaris*	w	V	V
American Bittern *Botaurus lentiginosus*	-	V	-
Least Bittern *Ixobrychus exilis*	-	V	-
Little Bittern *Ixobrychus minutus*	s	V	V
Night-heron *Nycticorax nycticorax*	s p	V	V
Yellow-crowned Night-heron *Nycticorax violaceus*	-	V	V
Green Heron *Butorides virescens*	-	V	V
Squacco Heron *Ardeola ralloides*	s p	V	V
Cattle Egret *Bubulcus ibis*	R	V	p
Little Blue Heron *Egretta caerulea*	-	V	-
Tricoloured Heron *Egretta tricolor*	-	V	-
Snowy Egret *Egretta thula*	-	V	-
Western Reef-egret *Egretta gularis*	V	V	V
Little Egret *Egretta garzetta*	R	w	r
Great White Egret *Egretta alba*	w	V	-
Grey Heron *Ardea cinerea*	r W	w	r
Great Blue Heron *Ardea herodias*	-	V	-
Purple Heron *Ardea purpurea*	s	V	p
Black Stork *Ciconia nigra*	s p w	V	V
White Stork *Ciconia ciconia*	R	V	V
Glossy Ibis *Plegadis falcinellus*	w	V	V
Spoonbill *Platalea leucorodia*	s p w	V	V
Greater Flamingo *Phoenicopterus roseus*	w p	V	-
Honey-buzzard *Pernis apivorus*	s p	-	V
American Swallow-tailed Kite *Elanoides forficatus*	-	V	-
Black-winged Kite *Elanus caeruleus*	r	-	-
Black Kite *Milvus migrans*	S P	V	p
Red Kite *Milvus milvus*	r w	V	V
White-tailed Eagle *Haliaeetus albicilla*	V	-	-
Lammergeier *Gypaetus barbatus*	V	-	-
Egyptian Vulture *Neophron percnopterus*	s p	V	V

SPECIES	Mainland Portugal	Azores	Madeira
Griffon Vulture *Gyps fulvus*	r p	-	-
Rueppell's Vulture *Gyps rueppellii*	V	-	-
Cinereous Vulture *Aegypius monachus*	r	-	-
Short-toed Eagle *Circaetus gallicus*	s p	-	-
Marsh Harrier *Circus aeruginosus*	r w	V	V
Hen Harrier *Circus cyaneus*	r w	V	-
Pallid Harrier *Circus macrourus*	V	-	-
Montagu's Harrier *Circus pygargus*	s p	V	V
Goshawk *Accipiter gentilis*	r	-	-
Sparrowhawk *Accipiter nisus*	r p	-	r
Buzzard *Buteo buteo*	R p	R	R
Long-legged Buzzard *Buteo rufinus*	V	-	V
Rough-legged Buzzard *Buteo lagopus*	-	V	-
Lesser Spotted Eagle *Aquila pomarina*	V	-	-
Spotted Eagle *Aquila clanga*	V	-	-
Spanish Imperial Eagle *Aquila adalberti*	r	-	-
Golden Eagle *Aquila chrysaetos*	r	-	-
Booted Eagle *Aquila pennata*	s p w	-	V
Bonelli's Eagle *Aquila fasciata*	r	-	-
Osprey *Pandion haliaetus*	p w	V	V
Lesser Kestrel *Falco naumanni*	s p	V	V
Kestrel *Falco tinnunculus*	R	V	R
American Kestrel *Falco sparverius*	-	V	-
Red-footed Falcon *Falco vespertinus*	V	V	V
Amur Falcon *Falco amurensis*	-	V	-
Merlin *Falco columbarius*	w	V	V
Hobby *Falco subbuteo*	s p	V	V
Eleonora's Falcon *Falco eleonorae*	p	V	V
Lanner Falcon *Falco biarmicus*	V	-	V
Saker Falcon *Falco cherrug*	V	-	-
Peregrine Falcon *Falco peregrinus*	r w	V	V
Barbary Falcon *Falco pelegrinoides*	-	-	V
Water Rail *Rallus aquaticus*	r	V	V
Spotted Crake *Porzana porzana*	p w	V	V
Sora *Porzana carolina*	-	V	V
Little Crake *Porzana parva*	V	V	V
Baillon's Crake *Porzana pusilla*	s?	V	V
Corncrake *Crex crex*	V	V	V
Moorhen *Gallinula chloropus*	R	r	r
Lesser Moorhen *Gallinula angulata*	-	-	V
Allen's Gallinule *Porphyrio alleni*	V	V	V
Purple Gallinule *Porphyrio martinicus*	V	V	V
Purple Swamphen *Porphyrio porphyrio*	R	-	V
Coot *Fulica atra*	r W	w	r
American Coot *Fulica americana*	V	V	V
Red-knobbed Coot *Fulica cristata*	V	-	-
Small Buttonquail *Turnix sylvaticus*	Ext	-	-
Crane *Grus grus*	w	V	V

SPECIES	Mainland Portugal	Azores	Madeira
Sandhill Crane *Grus canadensis*	-	V	-
Little Bustard *Tetrax tetrax*	r w	-	V
Great Bustard *Otis tarda*	r	-	-
Oystercatcher *Haematopus ostralegus*	w	V	V
African Oystercatcher *Haematopus moquini*	-	-	V
Black-winged Stilt *Himantopus himantopus*	S P w	V	V
Avocet *Recurvirostra avosetta*	r w	V	V
Stone-curlew *Burhinus oedicnemus*	r w	V	V
Cream-colored Courser *Cursorius cursor*	V	-	V
Collared Pratincole *Glareola pratincola*	s	-	V
Little Ringed Plover *Charadrius dubius*	p s w	V	p
Ringed Plover *Charadrius hiaticula*	W P	w	p
Semipalmated Plover *Charadrius semipalmatus*	-	V	-
Killdeer *Charadrius vociferus*	V	V	V
Kentish Plover *Charadrius alexandrinus*	R P	r	r
Lesser Sand Plover *Charadrius mongolus*	V	-	-
Dotterel *Charadrius morinellus*	p	V	V
Pacific Golden Plover *Pluvialis fulva*	V	V	V
American Golden Plover *Pluvialis dominica*	V	V	V
Golden Plover *Pluvialis apricaria*	W	V	V
Grey Plover *Pluvialis squatarola*	W P	w	V
Sociable Plover *Vanellus gregarius*	V	-	-
Lapwing *Vanellus vanellus*	r W	w	p
Red Knot *Calidris canutus*	w p	w p	p
Sanderling *Calidris alba*	W P	w	p
Semipalmated Sandpiper *Calidris pusilla*	V	V	V
Western Sandpiper *Calidris mauri*	-	V	V
Little Stint *Calidris minuta*	w P	V	p
Temminck's Stint *Calidris temminckii*	p	V	V
Least Sandpiper *Calidris minutilla*	-	V	-
White-rumped Sandpiper *Calidris fuscicollis*	V	V	V
Baird's Sandpiper *Calidris bairdii*	V	V	-
Pectoral Sandpiper *Calidris melanotos*	V	V	V
Sharp-tailed Sandpiper *Calidris acuminata*	V	V	V
Curlew Sandpiper *Calidris ferruginea*	p	V	p
Purple Sandpiper *Calidris maritima*	w	V	V
Dunlin *Calidris alpina*	W P	w p	p
Stilt Sandpiper *Calidris himantopus*	V	V	-
Broad-billed Sandpiper *Limicola falcinellus*	V	-	-
Buff-breasted Sandpiper *Tryngites subruficollis*	V	V	V
Ruff *Philomachus pugnax*	w p	V	p
Jack Snipe *Lymnocryptes minimus*	w	V	p
Common Snipe *Gallinago gallinago*	s P W	w r	w
Wilson's Snipe *Gallinago delicata*	V	w	-
Great Snipe *Gallinago media*	V	-	V
Short-billed Dowitcher *Limnodromus griseus*	-	V	-
Long-billed Dowitcher *Limnodromus scolopaceus*	V	V	-
Woodcock *Scolopax rusticola*	w	w r	r

SPECIES	Mainland Portugal	Azores	Madeira
Black-tailed Godwit *Limosa limosa*	W P	p	p
Hudsonian Godwit *Limosa haemastica*	-	V	-
Bar-tailed Godwit *Limosa lapponica*	w P	V	p
Whimbrel *Numenius phaeopus*	w P	w p	p
Slender-billed Curlew *Numenius tenuirostris*	V	V	-
Curlew *Numenius arquata*	W P	V	p
Upland Sandpiper *Bartramia longicauda*	V	V	V
Terek Sandpiper *Xenus cinereus*	V	V	-
Common Sandpiper *Actitis hypoleucos*	r P	V	p
Spotted Sandpiper *Actitis macularius*	V	V	V
Green Sandpiper *Tringa ochropus*	w p	V	V
Solitary Sandpiper *Tringa solitaria*	V	V	V
Spotted Redshank *Tringa erythropus*	w p	V	V
Greater Yellowlegs *Tringa melanoleuca*	V	V	-
Greenshank *Tringa nebularia*	w p	w p	p
Lesser Yellowlegs *Tringa flavipes*	V	V	V
Willet *Tringa semipalmata*	V	V	-
Marsh Sandpiper *Tringa stagnatilis*	V	V	-
Wood Sandpiper *Tringa glareola*	p	V	V
Redshank *Tringa totanus*	W P	V	p
Turnstone *Arenaria interpres*	W P	w p	p w
Wilson's Phalarope *Phalaropus tricolor*	V	V	V
Red-necked Phalarope *Phalaropus lobatus*	V	V	V
Grey Phalarope *Phalaropus fulicarius*	p	V	V
Pomarine Skua *Stercorarius pomarinus*	w p	V	p
Arctic Skua *Stercorarius parasiticus*	w p	V	p
Long-tailed Skua *Stercorarius longicaudus*	p	V	p
Great Skua *Stercorarius skua*	w p	V	p
South Polar Skua *Stercorarius maccormicki*	-	V	V
Great Black-headed Gull *Larus ichthyaetus*	-	-	V
Mediterranean Gull *Larus melanocephalus*	W P	V	p
Laughing Gull *Larus atricilla*	V	V	V
Franklin's Gull *Larus pipixcan*	V	V	V
Bonaparte's Gull *Larus philadelphia*	V	V	V
Black-headed Gull *Larus ridibundus*	r W P	w p	w p
Slender-billed Gull *Larus genei*	V	-	-
Audouin's Gull *Larus audouinii*	r p	V	-
Ring-billed Gull *Larus delawarensis*	w	w	w
Common Gull *Larus canus*	w	V	V
Lesser Black-backed Gull *Larus fuscus*	r P W	w p	W P
Herring Gull *Larus argentatus*	w	V	V
American Herring Gull *Larus smithsonianus*	V	V	V
Caspian Gull *Larus cachinnans*	V	-	-
Yellow-legged Gull *Larus michahellis*	R	W R	R
Iceland Gull *Larus glaucoides*	V	V	V
Glaucous Gull *Larus hyperboreus*	V	V	V
Great Black-backed Gull *Larus marinus*	w	V	V
Kittiwake *Rissa tridactyla*	w	w	w

SPECIES	Mainland Portugal	Azores	Madeira
Kelp Gull *Larus dominicanus*	V	-	-
Sabine's Gull *Xema sabini*	p	V	V
Little Gull *Hydrocoloeus minutus*	w p	V	V
Gull-billed Tern *Gelochelidon nilotica*	p	V	V
Caspian Tern *Hydroprogne caspia*	w p	V	V
Royal Tern *Sterna maxima*	V	V	-
Lesser Crested Tern *Sterna bengalensis*	V	-	-
Sandwich Tern *Sterna sandvicensis*	w P	p	w p
Roseate Tern *Sterna dougallii*	V	S	s
Common Tern *Sterna hirundo*	s P	S	S
Arctic Tern *Sterna paradisaea*	p	V	p
Forster's Tern *Sterna forsteri*	V	V	-
Bridled Tern *Onychoprion anaethetus*	-	V	-
Sooty Tern *Onychoprion fuscatus*	V	s	s
Little Tern *Sternula albifrons*	s P	V	V
Whiskered Tern *Chlidonias hybrida*	s p	V	V
Black Tern *Chlidonias niger*	p	V	V
White-winged Black Tern *Chlidonias leucopterus*	V	V	V
Brown Noddy *Anous stolidus*	-	V	-
Guillemot *Uria aalge*	r	-	-
Brünnich's Guillemot *Uria lomvia*	-	V	-
Razorbill *Alca torda*	W	V	-
Little Auk *Alle alle*	V	V	V
Puffin *Fratercula arctica*	w	V	V
Black-bellied Sandgrouse *Pterocles orientalis*	r	-	-
Pin-tailed Sandgrouse *Pterocles alchata*	r	-	-
Rock Dove *Columba livia*	r	r	R
Stock Dove *Columba oenas*	w p s	-	V
Woodpigeon *Columba palumbus*	R W	R	V
Madeira Laurel Pigeon *Columba trocaz*	-	-	r
Collared Dove *Streptopelia decaocto*	R	r	V
Turtle Dove *Streptopelia turtur*	s p	V	p
Mourning Dove *Zenaida macroura*	-	V	-
Great Spotted Cuckoo *Clamator glandarius*	s p	-	V
Cuckoo *Cuculus canorus*	S p	V	V
Black-billed Cuckoo *Coccyzus erythropthalmus*	-	V	-
Yellow-billed Cuckoo *Coccyzus americanus*	-	V	-
Barn Owl *Tyto alba*	R	V	r
Scops Owl *Otus scops*	s p	-	V
Eagle Owl *Bubo bubo*	r	-	-
Snowy Owl *Bubo scandiacus*	-	V	-
Little Owl *Athene noctua*	R	-	-
Tawny Owl *Strix aluco*	R	-	-
Long-eared Owl *Asio otus*	r	r	-
Short-eared Owl *Asio flammeus*	w	V	w p
Marsh Owl *Asio capensis*	V	-	-
Nightjar *Caprimulgus europaeus*	s p	V	V
Red-necked Nightjar *Caprimulgus ruficollis*	s p	-	V

SPECIES	Mainland Portugal	Azores	Madeira
Common Nighthawk *Chordeiles minor*	-	V	-
Chimney Swift *Chaetura pelagica*	V	V	-
Plain Swift *Apus unicolor*	-	-	S r
Swift *Apus apus*	S P	V	p
Pallid Swift *Apus pallidus*	S P	V	s
Alpine Swift *Apus melba*	s p	V	V
White-rumped Swift *Apus caffer*	s	-	-
Little Swift *Apus affinis*	V	V	V
Kingfisher *Alcedo atthis*	r p	V	V
Belted Kingfisher *Megaceryle alcion*	-	V	-
Bee-eater *Merops apiaster*	S P	V	p
Blue-cheeked Bee-eater *Merops persicus*	-	-	V
Roller *Coracias garrulus*	s p	V	p
Hoopoe *Upupa epops*	R	V	s
Wryneck *Jynx torquilla*	s p w	-	V
Northern Flicker *Colaptes auratus*	-	V	-
Iberian Green Woodpecker *Picus viridis (sharpei)*	r	-	-
Yellow-bellied Sapsucker *Sphyrapicus varius*	-	V	-
Great Spotted Woodpecker *Dendrocopos major*	R	-	-
Lesser Spotted Woodpecker *Dendrocopos minor*	r	-	-
Dupont's Lark *Chersophilus duponti*	Ext	-	-
Calandra Lark *Melanocorypha calandra*	r	-	V
Short-toed Lark *Calandrella brachydactyla*	s p	V	V
Lesser Short-toed Lark *Calandrella rufescens*	r	-	-
Crested Lark *Galerida cristata*	R	V	-
Thekla Lark *Galerida theklae*	R	-	-
Woodlark *Lullula arborea*	R	-	V
Skylark *Alauda arvensis*	r W	V	w p
Sand Martin *Riparia riparia*	S P	V	p
Tree Swallow *Tachycineta bicolor*	-	V	-
Purple Martin *Progne subis*	-	V	-
Crag Martin *Ptyonoprogne rupestris*	r w	-	V
Swallow *Hirundo rustica*	S P w	p	p
House Martin *Delichon urbicum*	S P	p	p
Red-rumped Swallow *Cecropis daurica*	s p	V	p
Cliff Swallow *Petrochelidon pyrrhonota*	-	V	-
Richard's Pipit *Anthus richardi*	w	-	-
Blyth's Pipit *Anthus godlewski*	V	-	-
Tawny Pipit *Anthus campestris*	s p w	V	V
Berthelot's Pipit *Anthus berthelotii*	-	-	r
Olive-backed Pipit *Anthus hodgsoni*	V	-	-
Tree Pipit *Anthus trivialis*	s P	V	V
Meadow Pipit *Anthus pratensis*	W	V	V
Red-throated Pipit *Anthus cervinus*	p	V	V
Buff-bellied Pipit *Anthus rubescens*	-	V	-
Water Pipit *Anthus spinoletta*	s W	-	-
Rock Pipit *Anthus petrosus*	w	-	-
Yellow Wagtail *Motacilla flava*	S P	V	V

SPECIES	Mainland Portugal	Azores	Madeira
Citrine Wagtail *Motacilla citreola*	V	V	V
Grey Wagtail *Motacilla cinerea*	r w	R	R
White Wagtail *Motacilla alba*	R W	V	w p
Cedar Waxwing *Bombycilla cedrorum*	-	V	-
Waxwing *Bombycilla garrulus*	V	V	-
Dipper *Cinclus cinclus*	r	-	-
Wren *Troglodytes troglodytes*	R	V	V
Grey Catbird *Dumetella carolinensis*	-	V	-
Dunnock *Prunella modularis*	r W	-	V
Alpine Accentor *Prunella collaris*	w	-	-
Rufous Bush Chat *Cercotrichas galactotes*	s p	-	V
Robin *Erithacus rubecula*	R P W	R	R
Nightingale *Luscinia megarhynchos*	S P	-	p
Bluethroat *Luscinia svecica*	w P	-	V
Red-flanked Bluetail *Tarsiger cyanurus*	V	-	-
Black Redstart *Phoenicurus ochruros*	R W	V	p
Redstart *Phoenicurus phoenicurus*	s p	V	p
Moussier's Redstart *Phoenicurus moussieri*	V	-	-
Whinchat *Saxicola rubetra*	s p	V	p
Siberian Stonechat *Saxicola maurus*	V	-	-
Stonechat *Saxicola rubicola*	R	V	V
Isabelline Wheatear *Oenanthe isabellina*	-	V	V
Wheatear *Oenanthe oenanthe*	s P	p	p
Black-eared Wheatear *Oenanthe hispanica*	s p	V	-
Desert Weather *Oenanthe deserti*	V	-	V
White-crowned Black Wheatear *Oenanthe leucopyga*	V	-	-
Black Wheatear *Oenanthe leucura*	r	-	-
Rock Thrush *Monticola saxatilis*	s p	-	V
Blue Rock Thrush *Monticola solitarius*	r	-	V
Wood Thrush *Hylocichla mustelina*	-	V	V
Hermit Thrush *Catharus guttatus*	-	V	-
Swainson's Thrush *Catharus ustulatus*	-	V	-
Grey-cheeked Thrush *Catharus minimus*	-	V	-
Ring Ouzel *Turdus torquatus*	w p	V	V
Blackbird *Turdus merula*	R	R	R
Eyebrowed Thrush *Turdus obscurus*	V	-	-
Naumann's Thrush *Turdus naumanni*	-	V	-
Red-throated Thrush *Turdus ruficollis*	-	-	V
Fieldfare *Turdus pilaris*	w	w	V
Song Thrush *Turdus philomelos*	r W	V	V
Redwing *Turdus iliacus*	W	V	w
Mistle Thrush *Turdus viscivorus*	r	V	V
American Robin *Turdus migratorius*	-	V	-
Cetti's Warbler *Cettia cetti*	R	-	-
Fan-tailed Warbler *Cisticola juncidis*	R	-	-
Grasshopper Warbler *Locustella naevia*	p	V	p
River Warbler *Locustella fluviatilis*	-	-	V

SPECIES	Mainland Portugal	Azores	Madeira
Savi's Warbler *Locustella luscinioides*	s	-	-
Western Olivaceous Warbler *Iduna opaca*	s	-	V
Moustached Warbler *Acrocephalus melanopogon*	V	-	-
Aquatic Warbler *Acrocephalus paludicola*	p	-	-
Sedge Warbler *Acrocephalus schoenobaenus*	p	V	p
Paddyfield Warbler *Acrocephalus agricola*	V	V	-
Reed Warbler *Acrocephalus scirpaceus*	S P	-	p
Great Reed Warbler *Acrocephalus arundinaceus*	S	-	V
Icterine Warbler *Hippolais icterina*	V	-	V
Melodious Warbler *Hippolais polyglotta*	S P	V	V
Dartford Warbler *Sylvia undata*	r	-	-
Spectacled Warbler *Sylvia conspicillata*	s p	-	r
Subalpine Warbler *Sylvia cantillans*	S p	-	V
Sardinian Warbler *Sylvia melanocephala*	R	-	V
Western Orphean Warbler *Sylvia hortensis*	s p	-	V
Lesser Whitethroat *Sylvia curruca*	V	-	V
Whitethroat *Sylvia communis*	s p	V	p
Garden Warbler *Sylvia borin*	s P	V	p
Blackcap *Sylvia atricapilla*	R W	R	R
Two-barred Warbler *Phylloscopus plumbeitarsus*	-	-	V
Arctic Warbler *Phylloscopus borealis*	V	-	-
Pallas's Warbler *Phylloscopus proregulus*	V	-	-
Yellow-browed Warbler *Phylloscopus inornatus*	V	V	V
Dusky Warbler *Phylloscopus fuscatus*	V	-	-
Western Bonelli's Warbler *Phylloscopus bonelli*	s p	-	V
Wood Warbler *Phylloscopus sibilatrix*	V	-	V
Chiffchaff *Phylloscopus collybita*	s P W	V	w p
Iberian Chiffchaff *Phylloscopus ibericus*	S p	-	V
Willow Warbler *Phylloscopus trochilus*	P	V	p
Ruby-crowned Kinglet *Regulus calendula*	-	V	-
Goldcrest *Regulus regulus*	w	R	-
Firecrest *Regulus ignicapilla*	r w	-	-
Madeira Kinglet *Regulus madeirensis*	-	-	R
Spotted Flycatcher *Muscicapa striata*	s P	V	p
Red-breasted Flycatcher *Ficedula parva*	V	V	V
Pied Flycatcher *Ficedula hypoleuca*	s P	V	p
Long-tailed Tit *Aegithalos caudatus*	R	-	-
European Crested Tit *Parus cristatus*	R	-	-
Coal Tit *Parus ater*	R	-	-
Blue Tit *Parus caeruleus*	R	-	-
Great Tit *Parus major*	R	-	-
Nuthatch *Sitta europaea*	R	-	-
Wallcreeper *Tichodroma muraria*	V	-	-
Short-toed Treecreeper *Certhia brachydactyla*	R	-	-
Penduline Tit *Remiz pendulinus*	s w	-	-
Golden Oriole *Oriolus oriolus*	S p	V	p
Red-backed Shrike *Lanius collurio*	s p	V	V
Southern Grey Shrike *Lanius meridionalis*	R	-	-

SPECIES	Mainland Portugal	Azores	Madeira
Woodchat Shrike *Lanius senator*	S P	V	p
Jay *Garrulus glandarius*	R	-	-
Azure-winged Magpie *Cyanopica cyanus*	R	-	-
Common Magpie *Pica pica*	R	-	-
Nutcracker *Nucifraga caryocatactes*	V	-	-
Chough *Pyrrhocorax pyrrhocorax*	r	-	-
Jackdaw *Corvus monedula*	r	V	V
Rook *Corvus frugilegus*	V	V	V
Carrion Crow *Corvus corone*	R	V	V
Raven *Corvus corax*	r	-	V
Spotless Starling *Sturnus unicolor*	R	-	V
Starling *Sturnus vulgaris*	W	R	w
Rose-coloured Starling *Pastor roseus*	V	-	V
House Sparrow *Passer domesticus*	R	-	V
Spanish Sparrow *Passer hispaniolensis*	R	-	r
Tree Sparrow *Passer montanus*	r p	-	-
Rock Sparrow *Petronia petronia*	r	V	r
White-winged Snowfinch *Montifringila nivalis*	V	-	-
White-eyed Vireo *Vireo griseus*	-	V	-
Yellow-throated Vireo *Vireo flavifrons*	-	V	-
Philadelphia Vireo *Vireo philadelphicus*	-	V	-
Red-eyed Vireo *Vireo olivaceus*	-	V	-
Chaffinch *Fringilla coelebs*	R W	R	R
Brambling *Fringilla montifringilla*	w	V	V
Serin *Serinus serinus*	R	V	V
Island Canary *Serinus canaria*	-	R	R
Citril Finch *Serinus citrinella*	V	-	-
Greenfinch *Carduelis chloris*	R	-	r
Goldfinch *Carduelis carduelis*	R W	-	r
Siskin *Carduelis spinus*	w	V	r
Linnet *Carduelis cannabina*	R	V	r
Twite *Carduelis flavirostris*	V	-	-
Common Redpoll *Carduelis flammea*	V	V	-
Arctic Redpoll *Carduelis hornemanni*	-	V	-
Common Crossbill *Loxia curvirostra*	r p w	V	V
Trumpeter Finch *Bucanetes githagineus*	V	-	-
Common Rosefinch *Carpodacus erythrinus*	V	-	-
Bullfinch *Pyrrhula pyrrhula*	r w	-	-
Azores Bullfinch *Pyrrhula murina*	-	r	-
Hawfinch *Coccothraustes coccothraustes*	r	-	V
Black-and-white Warbler *Mniotilta varia*	-	V	-
Blue-winged Warbler *Vermivora pinus*	-	V	-
Tennessee Warbler *Vermivora peregrina*	-	V	-
Northern Parula *Parula americana*	-	V	-
Yellow Warbler *Dendroica petechia*	-	V	V
Chestnut-sided Warbler *Dendroica pensylvanica*	-	V	-
Black-throated Blue Warbler *Dendroica caerulescens*	-	V	-
Black-throated Green Warbler *Dendroica virens*	-	V	-

SPECIES	Mainland Portugal	Azores	Madeira
Magnolia Warbler *Dendroica magnolia*	-	V	-
Yellow-rumped Warbler *Dendroica coronata*	-	V	-
Yellow-throated Warbler *Dendroica dominica*	-	V	-
Blackpoll Warbler *Dendroica striata*	-	V	-
American Redstart *Setophaga ruticilla*	-	V	V
Ovenbird *Seiurus aurocapilla*	-	V	-
Northern Waterthrush *Seiurus noveboracensis*	-	V	-
Common Yellowthroat *Geothlypis trichas*	V	V	-
Hooded Warbler *Wilsonia citrina*	-	V	-
Canada Warbler *Wilsonia canadensis*	-	V	-
Summer Tanager *Piranga rubra*	-	V	-
Scarlet Tanager *Piranga olivacea*	-	V	-
Savannah Sparrow *Passerculus sandwichensis*	-	V	-
Lincoln's Sparrow *Melospiza lincolni*	-	V	-
White-crowned Sparrow *Zonotrichia leucophrys*	-	V	-
White-throated Sparrow *Zonotrichia albicollis*	-	V	-
Dark-eyed Junco *Junco hyemalis*	-	V	-
Lapland Bunting *Calcarius lapponicus*	V	V	-
Snow Bunting *Plectrophenax nivalis*	w p	w	V
Yellowhammer *Emberiza citrinella*	r	-	-
Cirl Bunting *Emberiza cirlus*	R	-	-
Rock Bunting *Emberiza cia*	R	-	-
Ortolan Bunting *Emberiza hortulana*	s p	-	V
Rustic Bunting *Emberiza rustica*	V	-	-
Little Bunting *Emberiza pusilla*	V	-	-
Yellow-breasted Bunting *Emberiza aureola*	V	-	-
Reed Bunting *Emberiza schoeniclus*	r w	-	V
Pallas's Reed Bunting *Emberiza pallasi*	V	-	-
Corn Bunting *Emberiza calandra*	R	-	
Dickcissel *Spiza americana*	-	V	-
Rose-breasted Grosbeak *Pheucticus ludovicianus*	-	V	-
Blue Grosbeak *Guiraca caerulea*	-	V	-
Indigo Bunting *Passerina cyanea*	-	V	-
Bobolink *Dolichonyx oryzivorus*	-	V	-
Baltimore Oriole *Icterus galbula*	-	V	-

INTRODUCED SPECIES	Mainland Portugal	Azores	Madeira
Note: The species listed here comprise those belonging to 'category C' – i.e. non-native species that have been accidentally or deliberately introduced and that have formed self-sustaining wild populations			
Ruddy Duck *Oxyura jamaicensis*	V	V	-
Red-legged Partridge *Alectoris rufa*	-	r	r
Pheasant *Phasianus colchicus*	r	-	r
Ring-necked Parakeet *Psittacula krameri*	r	R	-
Crested Myna *Acridotheres cristatellus*	r	-	-
House Sparrow *Passer domesticus*	-	R	-

Black-headed Weaver *Ploceus melanocephalus*	r	-	-
Yellow-crowned Bishop *Euplectes afer*	r	-	r
Common Waxbill *Estrilda astrild*	R	r	-
Red Avadavat *Amandava amandava*	r	-	-
Greenfinch *Carduelis chloris*	-	r	-
Goldfinch *Carduelis carduelis*	-	R	-

The following species although mentioned in the text belong to 'category E', meaning that no self-sustaining populations are known so far

Blue-crowned Parakeet *Aratinga acuticaudata*	r		
Senegal Parrot *Poicephalus senegalus*	r		

SPECIES OF UNCERTAIN STATUS

Note: The species listed here comprise those belonging to 'category D' – i.e. species for which it is not certain if they have ever occurred in a wild state, either because birds involved in the records might have escaped from captivity or because the records are old and poorly documented. This category applies to the mainland only.

	Mainland Portugal		
Mute Swan *Cygnus olor*			
Canada Goose *Branta canadensis*			
Bar-headed Goose *Anser indicus*			
Falcated Duck *Anas falcata*			
Steller's Eider *Polysticta stelleri*			
Great White Pelican *Pelecanus onocrotalus*			
Pink-backed Pelican *Pelecanus rufescens*			
Marabou Stork *Leptoptilos crumenifer*			
Lesser Flamingo *Phoenicopterus minor*			
White-backed Vulture *Gyps africanus*			
Gyr Falcon *Falco rusticolus*			
Barbary Falcon *Falco pelegrinoides*			
Demoiselle Crane *Grus virgo*			
Houbara Bustard *Chlamydotis undulata*			
Ivory Gull *Pagophila eburnea*			
Rufous Turtle Dove *Streptopelia orientalis*			
Laughing Dove *Streptopelia senegalensis*			
Grey-faced Woodpecker *Picus canus*			
Black Woodpecker *Dryocopus martius*			
Middle Spotted Woodpecker *Dendrocopos medius*			
Thrush Nightingale *Luscinia luscinia*			
Marsh Warbler *Acrocephalus palustris*			
Marmora's Warbler *Sylvia sarda*			
Collared Flycatcher *Ficedula albicollis*			
Bearded Tit *Panurus biarmicus*			
Marsh Tit *Parus palustris*			
Lesser Grey Shrike *Lanius minor*			
Great Grey Shrike *Lanius excubitor*			

Yellow-billed Chough *Pyrrhocorax graculus*			
White-winged Crossbill *Loxia leucoptera*			
Parrot Crossbill *Loxia pityopsittacus*			
House Bunting *Emberiza sahari*			

AMPHIBIANS AND REPTILES LIST

Sources: Livro Vermelho dos Vertebrados de Portugal (2005) and Atlas dos Anfíbios e Répteis de Portugal (2010)

X – native

INT – introduced

SPECIES	Mainland Portugal	Azores	Madeira
Golden-striped Salamander *Chioglossa lusitanica*	X	-	-
Sharp-ribbed Newt *Pleurodeles waltl*	X	-	-
Fire Salamander *Salamandra salamandra*	X	-	-
Bosca's Newt *Triturus boscai*	X	-	-
Alpine Crested Newt *Triturus carnifex*	-	INT	-
Palmate Newt *Triturus helveticus*	X	-	-
Marbled Newt *Triturus marmoratus*	X	-	-
Iberian Midwife Toad *Alytes cisternasii*	X	-	-
Common Midwife Toad *Alytes obstetricans*	X	-	-
West Iberian Painted Frog *Discoglossus galganoi*	X	-	-
Western Spadefoot *Pelobates cultripes*	X	-	-
Parsley Frog *Pelodytes punctatus*	X	-	-
Iberian Parsley Frog *Pelodytes ibericus*	X	-	-
Common Toad *Bufo bufo*	X	-	-
Natterjack Toad *Bufo calamita*	X	-	-
Common Tree Frog *Hyla arborea*	X	-	-
Stripeless Tree Frog *Hyla meridionalis*	X	-	-
Iberian Frog *Rana iberica*	X	-	-
Green Frog *Rana perezi*	X	-	-
African Clawed Frog *Xenopus laevis*	INT	-	-
European Pond Terrapin *Emys orbicularis*	X	-	-
Spanish Terrapin *Mauremys leprosa*	X	-	-
Red-eared Slider *Trachemys scripta*	INT	-	-
Loggerhead Sea Turtle *Caretta caretta*	X	X	X
Green Sea Turtle *Chelonia mydas*	-	X	X
Hawksbill Turtle *Eretmochelys imbricata*	-	X	X
Kemp's Turtle *Lepidochelys kempii*	X	X	X
Leatherback Turtle *Dermochelys coriacea*	X	X	X
Turkish Gecko *Hemydactylus turcicus*	X	-	-
Amerafrican House Gecko *Hemidactylus mabouia*	-	-	INT
Selvagens Gecko *Tarentola bischoffi*	-	-	X
Moorish Gecko *Tarentola mauritanica*	X	-	-
Mediterranean Chameleon *Chamaeleo chamaeleon*	X	-	-
Slow Worm *Anguis fragilis*	X	-	-
Iberian Worm Lizard *Blanus cinereus*	X	-	-
Spiny-footed Lizard *Acanthodactylus erythrurus*	X	-	-
Ocellated Lizard *Lacerta lepida*	X	-	-

SPECIES	Mainland Portugal	Azores	Madeira
Iberian Rock Lizard *Lacerta monticola*	X	-	-
Schreiber's Green Lizard *Lacerta schreiberi*	X	-	-
Madeira Lizard *Lacerta dugesii*	-	-	X
Bocage's Wall Lizard *Podarcis bocagei*	X	-	-
Carbonell's Wall Lizard *Podarcis carbonelli*	X	-	-
Iberian Wall Lizard *Podarcis hispanica*	-	-	-
Italian Wall Lizard *Podarcis sicula*	INT	-	-
Large Psammodromus *Psammodromus algirus*	X	-	-
Spanish Psammodromus *Psammodromus hispanicus*	X	-	-
Bedriaga's Skink *Chalcides bedriagai*	X	-	-
Three-toed Skink *Chalcides striatus*	X	-	-
Horseshoe Whip Snake *Coluber hippocrepis*	X	-	-
Smooth Snake *Coronella austriaca*	X	-	-
Southern Smooth Snake *Coronella girondica*	X	-	-
Ladder Snake *Elaphe scalaris*	X	-	-
False Smooth Snake *Macroprotodon cucculatus*	X	-	-
Montpellier Snake *Malpolon monspessulanus*	X	-	-
Viperine Snake *Natrix maura*	X	-	-
Grass Snake *Natrix natrix*	X	-	-
Lataste's Viper *Vipera latastei*	X	-	-
Iberian Adder *Vipera seoanei*	X	-	-

MAMMALS LIST

Source: Livro Vermelho dos Vertebrados de Portugal (2005)
X – native
INT – introduced
EXT - extinct

SPECIES	Mainland Portugal	Azores	Madeira
European Hedgehog *Erinaceus europaeus*	X	INT	-
Eurasian Pygmy Shrew *Sorex minutus*	X	-	-
Iberian Shrew *Sorex granarius*	X	-	-
Mediterranean Water Shrew *Neomys anomalus*	X	-	-
Greater White-toothed Shrew *Crocidura russula*	X	-	-
Lesser White-toothed Shrew *Crocidura suaveolens*	X	-	-
Etruscan Shrew *Suncus etruscus*	X	-	-
Pyrenean Desman *Galemys pyrenaicus*	X	-	-
Iberian Mole *Talpa occidentalis*	X	-	-
Greater Horseshoe Bat *Rhinolophus ferrumequinum*	X	-	-
Lesser Horseshoe Bat *Rhinolophus hipposideros*	X	-	-
Mediterranean Horseshoe Bat *Rhinolophus euryale*	X	-	-
Mehely's Horseshoe Bat *Rhinolophus mehelyi*	X	-	-
Bechstein's Bat *Myotis bechsteini*	X	-	-
Mouse-eared Bat *Myotis myotis*	X	X	-
Lesser Mouse-Eared Bat *Myotis blythii*	X	-	-
Natterer's Bat *Myotis nattereri*	X	-	-
Geoffroy's Bat *Myotis emarginatus*	X	-	-

SPECIES	Mainland Portugal	Azores	Madeira
Whiskered Bat *Myotis mystacinus*	X	-	-
Daubenton's Bat *Myotis daubentonii*	X	-	-
Common Pipistrelle *Pipistrellus pipistrellus*	X	-	-
Nathusius' Pipistrelle *Pipistrellus nathusii*	X	-	-
Kuhl's Pipistrelle *Pipistrellus kuhlii*	X	-	-
Madeira Pipistrelle *Pipistrellus maderensis*	-	X	X
Soprano Pipistrelle *Pipistrellus pygmaeus*	X	-	-
Savi's Pipistrelle *Hypsugo savii*	X	-	-
Lesser Noctule *Nyctalus leisleri*	X	-	X
Azores Noctule *Nyctalus azoreum*	-	X	-
Common Noctule *Nyctalus noctula*	X	-	-
Greater Noctule Bat *Nyctalus lasiopterus*	X	-	-
Serotine Bat *Eptesicus serotinus*	X	-	-
Western Barbastelle *Barbastella barbastellus*	X	-	-
Brown Long-eared Bat *Plecotus auritus*	X	-	-
Grey Long-eared Bat *Plecotus austriacus*	X	-	X
Schreiber's Bat *Miniopterus schreibersi*	X	-	-
European Free-tailed Bat *Tadarida teniotis*	X	-	X
European Rabbit *Oryctolagus cuniculus*	X	INT	INT
Iberian Hare *Lepus granatensis*	X	-	-
Eurasian Red Squirrel *Sciurus vulgaris*	X	-	-
European Water Vole *Arvicola terrestris*	X	-	-
Southwestern Water Vole *Arvicola sapidus*	X	-	-
Cabrera's Vole *Microtus cabrerae*	X	-	-
Common Vole *Microtus arvalis*	X	-	-
Field Vole *Microtus agrestis*	X	-	-
Lusitanian Pine Vole *Microtus lusitanicus*	X	-	-
Mediterranean Pine Vole *Microtus duodecimcostatus*	X	-	-
Wood Mouse *Apodemus sylvaticus*	X	-	-
Black Rat *Rattus rattus*	X	INT	INT
Brown Rat *Rattus norvegicus*	X	INT	INT
House Mouse *Mus domesticus*	X	INT	INT
Western Mediterranean Mouse *Mus spretus*	X	INT	-
Garden Dormouse *Eliomys quercinus*	X	-	-
Rough-toothed Dolphins *Steno bredanensis*	-	X	X
Bottlenose Dolphin *Tursiopus truncatus*	X	X	X
Striped Dolphin *Stenella coeruleoalba*	X	X	X
Atlantic Spotted Dolphin *Stenella frontalis*	-	X	X
Common Dolphin *Delphinus delphis*	X	X	X
Risso's Dolphin *Grampus griseus*	X	X	X
False Killer Whale *Pseudorca crassidens*	X	X	X
Long-finned Pilot Whale *Globicephala melaena*	X	-	X
Short-finned Pilot Whale *Globicephala macrorhynchus*	-	X	X
Killer Whale *Orcinus orca*	X	X	X
Harbour Porpoise *Phocoena phocoena*	X	-	-
Cuvier's Beaked Whale *Ziphius cavirostris*	X	X	X
Northern Bottlenose Whale *Hyperoodon ampullatus*	-	X	-

SPECIES	Mainland Portugal	Azores	Madeira
True's Beaked Whale *Mesoplodon mirus*	-	X	X
Gervais' Beaked Whale *Mesoplodon europaeus*	-	X	X
Sowerby's Beaked Whale *Mesoplodon bidens*	-	X	X
Blainville's Beaked Whale *Mesoplodon densirostris*	-	X	X
Pygmy Sperm Whale *Kogia breviceps*	X	X	X
Sperm Whale *Physeter macrocephalus*	X	X	X
Common Minke Whale *Balaenoptera acutorostrata*	X	X	X
Sei Whale *Balaenoptera borealis*	X	X	X
Fin Whale *Balaenoptera physalus*	X	X	X
Blue Whale *Balaenoptera musculus*	X	X	X
Humpback Whale *Megaptera novaeangliae*	X	X	X
North Atlantic Right Whale *Eubalaena glacialis*	X	X	X
Red Fox *Vulpes vulpes*	X	-	-
Grey Wolf *Canis lupus*	X	-	-
Brown Bear *Ursus arctos*	EXT	-	-
Stoat *Mustela erminea*	X	-	-
Least Weasel *Mustela nivalis*	X	INT	-
American Mink *Mustela vison*	INT	-	-
European Polecat *Mustela putorius*	X	-	-
Beech Marten *Martes foina*	X	-	-
European Pine Marten *Martes martes*	X	-	-
European Badger *Meles meles*	X	-	-
Common Otter *Lutra lutra*	X	-	-
Common Genet *Genetta genetta*	INT	-	-
Egyptian Mongoose *Herpestes ichneumon*	INT	-	-
Wildcat *Felis silvestris*	X	-	-
Iberian Lynx *Lynx pardinus*	X	-	-
Mediterranean Monk Seal *Monachus monachus*	-	-	X
Bearded Seal *Erignatus barbatus*	X	-	-
Hooded Seal *Cystophora cristata*	X	-	-
Grey Seal *Halichoerus grypus*	X	-	-
Harbour Seal *Phoca vitulina*	X	-	-
Ringed Seal *Phoca hispida*	X	-	-
Wild Boar *Sus scrofa*	X	-	-
Fallow Deer *Dama dama*	INT	-	-
Red Deer *Cervus elaphus*	X	-	-
European Roe Deer *Capreolus capreolus*	X	-	-
Argali *Ovis ammon*	INT	-	-
Iberian Wild Goat *Capra pyrenaica*	X	-	-

RECOMMENDED READING

General Works

Cabral, M. J. (coord.), Almeida, J., Almeida, P. R., Dellinger, T., Ferrand de Almeida, N., Oliveira, M. E., Palmeirim, J. M., Queiroz, A. L., Rogado, L. & Santos Reis, M. (eds) 2005. *Livro Vermelho dos Vertebrados de Portugal*. Instituto da Conservação da Natureza. Lisboa. [The Red Data Book on vertebrates]

Costa, H., De Juana, E. & Varela, J. 2011. *Aves de Portugal, incluindo os arquipélagos dos Açores, da Madeira e das Selvagens*. Lynx Edicións, Barcelona. [Field guide to the birds of Portugal, including Azores, Madeira and Selvagens]

Costa, L. T., Nunes, M., Geraldes, P. & Costa, H. 2003. *Zonas Importantes para as Aves em Portugal*. SPEA-Sociedade Portuguesa para o Estudo das Aves, Lisboa. [Important Bird Areas in Portugal]

Crochet, P.-A., Barthel, P. H., Bauer, H.-G., van den Berg, A. B., Bezzel, E., Collinson, J. M., Dietzen, C., Dubois, P. J., Fromholtz, J., Helbig, A. J. †, Jiguet, F., Jirle, E., Knox, A.G., Krüger, T., Le Maréchal, P., van Loon, A.J., Päckert, M., Parkin, D. T., Pons, J.-M., Raty, L., Roselaar, C. S., Sangster, G., Steinheimer, F. D., Svensson, L., Tyrberg, T., Votier, S. C. & Yésou, P. 2011 *AERC TAC's taxonomic recommendations: 2011 report*. Available online at http://www.aerc.eu/tac.html. Viewed 30 October 2013

Crochet, P.-A., Barthel, P. H., Bauer, H.-G., van den Berg, A. B., Bezzel, E., Collinson, J. M., Dietzen, C., Dubois, P. J., Fromholtz, J., Helbig, A. J. †, Jiguet, F., Jirle, E., Knox, A.G., Krüger, T., Le Maréchal, P., van Loon, A.J., Päckert, M., Parkin, D. T., Pons, J.-M., Raty, L., Roselaar, C. S., Sangster, G., Steinheimer, F. D., Svensson, L., Tyrberg, T., Votier, S. C. & Yésou, P. 2012 *AERC TAC's taxonomic recommendations: 2012 report*. Available online at http://www.aerc.eu/tac.html. Viewed 30 October 2013

Crochet P.-A. & Joynt G. 2012. *AERC list of Western Palearctic birds*. December 2012 version. Available online at http://www.aerc.eu/tac.html. Viewed 21 March 2013

Ramírez, I., Geraldes, P., Meirinho, A., Amorim, P. & Paiva, V. 2008. *Áreas Importantes para as Aves Marinhas em Portugal*. SPEA, Lisboa. [Important seabird areas in Portugal]

Mainland Portugal

Barros, F. & Marques, P. 1999. *Atlas das Aves que Nidificam na Serra de Montejunto*. ADSAICA-Associação de Desenvolvimento das Serras de Aire e Candeeiros. [An atlas of the breeding birds of serra de Montejunto]

BirdLife International 2004. *Birds in Europe: Population Estimates, Trends and Conservation Status*. Cambridge UK: BirdLife International, BirdLife Conservation Series nº 12.

Bolton, M. 1987. *An Atlas of Wintering Birds in the Western Algarve.* Occasional Publication nº 1. A Rocha, Portimão.

Catry, P., Costa, H., Elias, G. & Matias, R. 2010. *Aves de Portugal. Ornitologia do Território Continental.* Assírio & Alvim, Lisboa. [The birds of Mainland Portugal]

Costa, H. 2003. *Onde Observar Aves no Sul de Portugal.* Assírio & Alvim, Lisboa. [Where to watch birds in Southern Portugal]

Elias, G., Reino, L. M., Silva, T., Tomé, R. & Geraldes, P. 1999. *Atlas das Aves Invernantes do Baixo Alentejo.* SPEA-Sociedade Portuguesa para o Estudo das Aves, Lisboa. [An atlas of the wintering birds of Baixo Alentejo]

Elias, G., Costa, H., Moore, C. C. & Franco, C. 2006. *Aves do Estuário do Sado.* Instituto da Conservação da Natureza/Reserva Natural do Estuário do Sado. [The birds of Sado estuary]

Equipa Atlas 2008. *Atlas das Aves Nidificantes em Portugal (1999-2005).* Instituto da Conservação da Natureza, Sociedade Portuguesa para o Estudo das Aves, Parque Natural da Madeira e Secretaria Regional do Ambiente e do Mar. Assírio & Alvim, Lisboa. [An atlas of the breeding birds of Portugal]

Leitão, D., Catry, P. Costa, H., Elias, G. & Reino, L. 1998. *As Aves do Estuário do Tejo.* Instituto da Conservação da Natureza, Lisboa. [The birds of the Tagus estuary]

Loureiro, A., Ferrand de Almeida, N., Carretero, M. A. & Paulo, O. S. (eds) 2010. *Atlas dos Anfíbios e Répteis de Portugal.* Esfera do Caos, Lisboa. [An atlas of the amphibians and reptiles of Portugal]

Matias, R. 2002. *Aves Exóticas que Nidificam em Portugal Continental.* Instituto da Conservação da Natureza, Lisboa. [Field Guide to non-native breeding birds in mainland Portugal]

Matias, R., Catry, P., Costa, H., Elias, G., Jara, J, Moore, C. C. & Tomé, R. 2007. Lista sistemática das aves de Portugal Continental. *Anuário Ornitológico* 5: 74-132. Available on line at http://www.spea.pt/fotos/editor2/lista_avesptc_anuario5.pdf [Systematic list of the birds of mainland Portugal]. Viewed 30 October 2013.

Matias, R., Alfrey, P., Costa, H., Jara, J., Moore, C. C., Santos, J. L. & Tipper, R. 2011. Adições e alterações à lista das aves de Portugal Continental: primeira actualização. *Anuário Ornitológico* 8: 105-117. Available on line at http://www.spea.pt/fotos/editor2/anuario_ornitologico8_4.pdf [Additions and changes to the systematic list of the birds of Mainland Portugal: first update]. Viewed 30 October 2013.

Moore, C. C., Elias, G. & Costa, H. 1997. Seawatching off Portugal. *Birding World* 10 (11): 433-434.

Moore, C. C. 2001. *Berlenga Bird Year*. Reserva Natural das Berlengas, ICN.

Moore, C. C. & Matias, R. 2007. Removal of Ménétries Warbler *Sylvia mystacea* from the list of the Portuguese avifauna. *Anuário Ornitológico* 5: 140-143.

Pimenta, M. & Santarém, L. 1996. *Atlas das Aves do Parque Nacional da Peneda-Gerês*. Instituto da Conservação da Natureza e Parque Nacional da Peneda-Gerês. [An atlas of the birds of Peneda-Gerês National Park]

Azores

Bannerman, D. A. & Bannerman W. M. 1966. *Birds of the Atlantic Islands. Vol. III. A History of the Birds of the Azores.* Oliver & Boyd, Edinburgh and London.

Borges, P. A.V., Costa, A., Cunha, R., Gabriel, R., Gonçalves, V., Martins, A. F., Melo, I., Parente, M., Raposeiro, P., Rodrigues, P., Santos, R. S., Silva, L., Vieira, P. & Vieira, V. (eds) 2010. *A List of the Terrestrial and Marine Biota from the Azores*. Princípia, Cascais.

Clarke, T. 2006. *Birds of the Atlantic Islands*. Christopher Helm, London.

Le Grand, G. 1983. Check list of the birds of the Azores. *Arquipélago Serv. Ciências Nat.* 4: 49-58.

Monteiro, L. R, Ramos, J. A., Pereira, J. C., Monteiro, P. R., Feio R. S., Thompson, D. R., Bearshop, S., Furness, R. W., Laranjo, M., Hilton, G., Neves, V. C., Groz, M. P., Thompson, K. C. 1999. Status and distribution of Fea's Petrel, Bulwer's Petrel, Manx Shearwater, Little Shearwater and Band-rumped Storm-petrel in the Azores. *Waterbirds* 22: 358-366.

Pereira, C. 2010. *Aves dos Açores*. SPEA-Sociedade Portuguesa para o Estudo das Aves, Lisboa. [An identification guide to the birds of the Azores]

Rodebrand, S. (comp.) 2012. *Checklist of the birds of the Azores including 2011.* Available online at: http://www.birdingazores.com/?page=2011checklist. Viewed 21 March 2013.

Madeira

Bannerman, D. A. 1963. *Birds of the Atlantic Islands. Vol. I. A History of the Birds of the Canary Islands and the Selvages.* Oliver & Boyd, Edinburgh and London.

Bannerman, D. A. & Bannerman W. M. 1965. *Birds of the Atlantic Islands. Vol. II. A History of the Birds of Madeira, the Desertas and the Porto Santo Islands.* Oliver & Boyd, Edinburgh and London.

Biscoito, M. & Zino, F. 2002. *Aves do Arquipélago da Madeira.* Biodiversidade Madeirense: Avaliação e Conservação n° 3. Secretaria Regional do Ambiente e dos Recursos Naturais, Funchal. [The birds of Madeira]

Borges, P. A.V., Abreu, C., Aguiar, A. M. F., Carvalho, P., Jardim, R., Melo, I., Oliveira, P., Sérgio, C., Serrano, A. R. M. & Vieira, P. (eds) 2008. *A List of the Terrestrial Fungi, Flora and Fauna of Madeira and Selvagens Archipelagos.* Direcção Regional do Ambiente da Madeira e Universidade dos Açores, Funchal & Angra do Heroísmo.

Câmara, D. 1997. *Guia de Campo das Aves do Parque Ecológico do Funchal e do Arquipélago da Madeira.* Associação dos Amigos do Parque Ecológico, Funchal. [A fieldguide to birds of Parque Ecológico do Funchal and the Madeira archipelago]

Clarke, T. 2006. *Birds of the Atlantic Islands.* Christopher Helm, London.

Delgado, C. 2006. *Onde Observar Aves no Arquipélago da Madeira.* SPEA-Sociedade Portuguesa para o Estudo das Aves, Funchal. [Where to watch birds in Madeira]

Mougin, J-L., Roux, F., Zino, P. A., Jouanin, C., Stahl, J.-C. & Despin, B. 1987. Les oiseaux visiteurs des Iles Selvagens. *Bol. Mus. Mun. Funchal* 39: 5-24. [The birds of Selvagens]

Oliveira, P. & Menezes, D. 2004. *Aves do Arquipélago da Madeira.* Serviço do Parque Natural da Madeira e Arquipélago Verde Produtos Pomocionais Lda. Funchal. [The birds of Madeira]

Romano, H., Correia-Fagundes, C., Zino, F. & Biscoito, M. J. 2010. Birds of the archipelago of Madeira and the Selvagens. II – new records and checklist update. *Bol. Mus. Mun. Funchal* 60 (326): 5-44.

Zino, F., Biscoito, M. J., Zino, P.A. 1995. Birds of the archipelago of Madeira and the Selvagens new records and checklist. *Bol. Mus. Mun. Funchal* 47 (262): 63-100.

USEFUL CONTACTS AND WEBSITES

General tourist information and logistics

NAME AND DESCRIPTION	WEB ADDRESS	TELEPHONE NR. (COUNTRY CODE 351)
Mainland Portugal		
Visit Portugal	http://www.visitportugal.com	
Visit Porto and North	http://visitportoandnorth.travel	
Visit Centro	http://www.visitcentro.com/en	
Lisbon: Visit Lisboa	http://www.visitlisboa.com	
Visit Alentejo	http://www.visitalentejo.pt	
Visit Algarve	http://www.visitalgarve.pt	
Airports (flights)	http://www.ana.pt	Airport contacts: Lisbon: 21 841 35 00; Algarve: 289 800 800; Porto: 22 943 24 00
Motorway maps	http://www.highwaymaps.eu/portugal	
Tolls	http://www.portugaltolls.pt	
Petrol prices	http://www.precoscombustiveis.dgeg.pt	
CP (trains)	http://www.cp.pt	707 201 280
Rede Expressos (buses)	http://www.rede-expressos.pt	707 223 344
Santos (buses to the north-east)	http://www.santosviagensturismo.pt	279 652 188
Tide table	http://www.hidrografico.pt	
Automóvel Club de Portugal (maps)	http://www.acp.pt	21 318 01 00
Instituto Geográfico do Exército (maps)	http://www.igeoe.pt	21 850 53 00
Instituto Geográfico Português (maps)	http://www.igeo.pt	21 381 96 00
Embassies and consulates	http://www.oi.acidi.gov.pt/modules.php?name=Content&pa=showpage&pid=10	
On-duty pharmacies	http://www.farmaciasdeservico.net	
Birding		
Aves de Portugal	http://www.avesdeportugal.info	
Reservoir Birds	http://www.reservoirbirds.es/index.asp	
Associação Portuguesa de Anilhadores de Aves (APAA – an association of Portuguese bird ringers)	http://apaa.pt/apaaenglish1.html	
Comité Português de Raridades (PRC – the Portuguese Rarities Committee)	http://www.spea.pt/pt/observar-aves/comite-portugues-raridades	

State bodies and protected areas	Instituto da Conservação da Natureza e das Florestas (ICNF – state body related with nature conservation)	http://www.icnf.pt/cn/ICNPortal/vEN2007	21 3507 900
Non-governmental organisations	Sociedade Portuguesa para o Estudo das Aves (SPEA)	http://www.spea.pt	21 322 04 30
	Almargem (a local NGO in Algarve)	http://almargem.org	289 412 959
	A Rocha (a local NGO in Algarve)	http://www.arocha.org/pt-pt	282 968 380
Site information: Mainland	Minho estuary: ferries	http://www.cm-caminha.pt ver.php?cod=1E0A	258 710 300
	Porto: underground	http://www.metrodoporto.pt/en	
	Ria de Aveiro: boats	http://www.moveaveiro.pt/03horarios/moveferryh.htm	234 380 560
	Ria de Aveiro: Bioria	http://www.bioria.com	234 840 600
	Ribacoa: Faia Brava	http://atnatureza.blogspot.pt	
		http://www.atnatureza.org	271 311 202
	Serra da Estrela: CISE – Centro de Interpretação da Serra da Estrela	http://www.cise.pt/pt	238 320 300
	Berlengas: boats	http://www.viamar-berlenga.com	262 785 646
	Tagus estuary: Associação de Beneficiários da Lezíria Grande (access to Ponta da Erva plains)	http://www.ablgvfx.pt	263 274 214
	Tagus estuary: salinas do Samouco	http://www.salinasdosamouco.pt	
	Tagus estuary: trains	http://www.fertagus.pt	707 127 127
	Tagus estuary: buses	http://www.tsuldotejo.pt	
	Tagus estuary: boats	http://www.transtejo.pt	
	Lisbon city: buses	http://www.carris.pt/en/home	
	Lisbon city: underground	http://www.metrolisboa.pt/eng	
	Costa do Estoril and Sintra: local buses	http://www.scotturb.com	21 469 91 25
	Lisbon region: bird sightings	http://www.birdinglisbon.com	
	Lagoa de Albufeira: Interpretation centre	http://sesimbra.pt/lagoapequena	93 998 22 92
	Serra da Arrábida: convent	http://www.foriente.pt/68/visitas-e-contactos.htm	21 219 7620

	Sado estuary: boats to Tróia	http://www.atlanticferries.pt	265 235 101
	Castro Verde: CEAVG – interpretation centre Vale Gonçalinho	http://www.lpn.pt	286 328 309
	Ria Formosa: boats	http://www.olhao.web.pt/horariobarcos.htm and http://www.cm-faro.pt/421/barcos-para-as-ilhas.aspx	
	Azores		
General information and logistics	Visit Azores	http://www.visitazores.com	
	Airports (flights)	http://www.ana.pt	
	SATA (inter-island flights)	http://www.sata.pt	
	Boats (Faial – Pico – São Jorge)	http://www.transmacor.pt	292 200 380
	Boats	http://www.atlanticoline.pt	800 500 501
	Direcção Regional de Ambiente dos Açores (DRAA – regional body related with environment proteccion)	http://servicos.sram.azores.gov.pt/doit	296 206 700
Birding	Aves dos Açores	http://azores.avesdeportugal.info	
	Birding Azores	http://www.birdingazores.com	
	Priolo's Environmental Centre	http://centropriolo.spea.pt	918 536 123
	Azores Bird Sightings	http://azoresbs.weebly.com/blog.html	
	Madeira		
General information and logistics	Visit Madeira	http://www.visitmadeira.pt	291 214 360
	Madeira airports (flights)	http://www.anam.pt/madeira_en	291 520 700
	Boats (Madeira to Porto Santo)	http://www.portosantoline.pt	291 210 300
	SATA (flights to Porto Santo)	http://www.sata.pt	
	Parque Natural da Madeira (Madeira Natural Park)	http://www.pnm.pt	
Birding	Birding Madeira	http://madeira.seawatching.net	
	SPEA - Madeira	http://www.spea.pt	291 214 360

www.ingramcontent.com/pod-product-compliance
Lightning Source LLC
Chambersburg PA
CBHW062136280125
20968CB00003B/37
* 9 7 8 1 7 8 4 2 7 2 2 3 4 *